Milk Protein - New Research Approaches

Edited by Narongsak Chaiyabutr

Published in London, United Kingdom

Milk Protein - New Research Approaches
http://dx.doi.org/10.5772/intechopen.98158
Edited by Narongsak Chaiyabutr

Contributors
Imen Fguiri, Manel Ziadi, Amel Sboui, Naziha Ayeb, Samira Arroum, Touhami Khorchani, Moufida Atigui, Mohamed Hammadi, Ameae M. Walker, Wei Wu, Changhui Deng, Jennifer L. Brockman, Linda A. Schuler, Morakot Nuntapaitoon, Dengpan Bu, Lu Ma, Zitai Guo, Roua Lajnaf, Mohamed Ali Ayadi, Hamadi Attia, Sawsan Feki, Hatem Masmoudi, Dasarapu Indrani, Jagatheesan Alagesan, Prathap Suganthirababu, M.V Sowmya, Dubba NagaRaju, Narongsak Chaiyabutr, Sumpun Thammacharoen, Nungnuch Saipin, Thiet Nguyen

First published in London, United Kingdom, 2022 by IntechOpen
IntechOpen is the global imprint of INTECHOPEN LIMITED, registered in England and Wales, registration number: 11086078, 5 Princes Gate Court, London, SW7 2QJ, United Kingdom

British Library Cataloguing-in-Publication Data
A catalogue record for this book is available from the British Library

Additional hard and PDF copies can be obtained from orders@intechopen.com

Milk Protein - New Research Approaches
Edited by Narongsak Chaiyabutr
p. cm.
Print ISBN 978-1-80355-201-9
Online ISBN 978-1-80355-202-6
eBook (PDF) ISBN 978-1-80355-203-3

For EU product safety concerns:
IN TECH d.o.o., Prolaz Marije Krucifikse Kozulić 3, 51000 Rijeka, Croatia,
info@intechopen.com or visit our website at intechopen.com.

Meet the editor

Narongsak Chaiyabutr is Professor Emeritus of Veterinary Physiology, Chulalongkorn University, Thailand. He obtained his DVM from the same university, and an MS in Physiology from Mahidol University, Thailand. He completed postgraduate training in Animal Physiology and Biochemistry under an FAO fellowship at the Royal Veterinary and Agricultural University, Denmark. He was awarded a Ph.D. with his research study at the Hannah Research Institute, Glasgow University, Scotland, UK. His current research is centered on the study of the control mechanism of mammary function during exposure to high ambient temperature in both dairy goats and crossbred dairy cattle in the tropics, relating the utilization of glucose for the synthesis of milk components in the mammary gland.

Contents

Preface **IX**

Section 1
Introduction **1**

Chapter 1 3
Introductory Chapter: Milk Protein Synthesis, Progress, and Projections
by Narongsak Chaiyabutr

Section 2
Microbiological and Biochemical Characteristics of Milk Production **11**

Chapter 2 13
Microbiological Quality and Biochemical Characteristics of Lactic Acid Bacteria from Camel Milk as Affected by the Production System and Stage of Lactation
by Imen Fguiri, Manel Ziadi, Amel Sboui, Naziha Ayeb, Moufida Atigui, Samira Arroum, Mohamed Hammadi and Touhami Khorchani

Section 3
Cellular and Molecular Milk Protein Synthesis **23**

Chapter 3 25
Superior Stimulation of β-Casein mRNA Accumulation by Pseudophosphorylated Prolactin: Enhanced Transcription and Message Stabilization
by Wei Wu, Changhui Deng, Jennifer L. Brockman, Linda A. Schuler and Ameae M. Walker

Chapter 4 37
Omics, the New Technological Approaches to the Milk Protein Researches
by Zitai Guo, Lu Ma and Dengpan Bu

Section 4
Milk Protein and Food Allergens **57**

Chapter 5 **59**
Characteristics of Cow Milk Proteins and the Effect of Processing on Their Allergenicity
by Roua Lajnaf, Sawsan Feki, Hamadi Attia, Mohamed Ali Ayadi and Hatem Masmoudi

Section 5
Control Mechanism of Breast Milk **79**

Chapter 6 **81**
Compare the Effects of Ultrasound versus Taping in Lactating Mothers with Breast Engorgement
by Dasarapu Indrani, Jagatheesan Alagesan, Prathap Suganthirababu, M.V. Sowmya and Dubba NagaRaju

Section 6
The Colostrum and Milk **91**

Chapter 7 **93**
Colostrum and Milk in Sow
by Morakot Nuntapaitoon

Section 7
High Ambient Temperature and Milk Protein Biosythesis **115**

Chapter 8 **117**
Effects of High Ambient Temperature on Milk Protein Synthesis in Dairy Cows and Goats: Insights from the Molecular Mechanism Studies
by Sumpun Thammacharoen, Nungnuch Saipin, Thiet Nguyen and Narongsak Chaiyabutr

Preface

Milk protein synthesis during lactation is a complex biological activity that involves diverse and dynamic interactions between proteins, genes, and other factors. The importance of milk protein to the neonate derives from its nutritional and immunological properties. The wide diversity in milk protein content of different species has prompted investigations into the source of these variations and milk protein's potential biological functions. *Milk Protein - New Research Approaches* presents comprehensive information on factors that might control milk protein production as well as discusses the genetic, cellular, and molecular progress being made in the field of dairy science. Chapters present recent and relevant research in the field of milk protein production and include extensive bibliographies. This volume is a useful resource for students, researchers, and professionals in veterinary, dairy, food, and animal science, among others.

Narongsak Chaiyabutr
Faculty of Veterinary Science,
Chulalongkorn University,
Bangkok, Thailand

Section 1

Introduction

Chapter 1

Introductory Chapter: Milk Protein Synthesis, Progress, and Projections

Narongsak Chaiyabutr

1. Introduction

Milk protein is one of the most important nutrients to the neonate drives their nutritional and immunological properties. It contains a variety of essential amino acids required by the body to maintain a variety of potential biological functions. Milk proteins fall into the category in which they are synthesized by the cells of the mammary gland. Milk protein is synthesized by precursors from the diet and the availability of amino acids (AA) passes from the blood to the lumen of the mammary alveoli. It is necessary to consider the basis for the two routes of the passage of specific proteins across the mammary epithelial cell [1, 2]. The first is the paracellular route that proteins pass from the extracellular fluid (ECF) to the lumen of the alveoli between adjacent secretory cells. The second is the transcellular route, which is proteins across the apical secretory cell membrane. The sequence of a series of steps viz. (i) the passage of amino acids from the blood capillary to the extracellular fluid (ECF), (ii) passage from the ECF to the secretory cell, (iii) the intracellular synthesis of milk specific proteins, (iv) intracellular translocation of the protein to the apical membrane of the mammary cell, (v) the protein across the apical membrane into the alveolar lumen. Therefore, the process necessitates the factors controlling each step, which are involved in a process of controlled polymerization within the mammary cell. The current knowledge of the mechanisms involved in these separate steps.

The knowledge in the area of milk protein continues to be one of rapid progress in the subject of lactation. Several papers review the finding involving milk protein in detail in this area. Since, the process of milk protein synthesis requires the accumulation of information from the testing of hypotheses in a wide range of species. New research approaches have provided a further understanding of the control mechanism of milk protein synthesis at both cellular and molecular levels. The process of the mammary gland and milk protein secretion impinge directly on many diverse topics such as sociology, medicine involving breast and food allergens feeding, agriculture involving the dairy and pig industry, and economic, etc.

2. Milk protein and area of research

Since the synthesis of milk protein during lactation is a complex biological activity. The mechanisms of milk protein synthesis are concerned with the mechanisms

IntechOpen

of other major milk constituents with lactose and milk fat which are widespread and diverse. Multifactorial have involved these lines of approach for the research area in both cellular or molecular studies of milk protein synthesis still being investigated by scientists. The mammary gland is the target organ of a complex system to the major pathway that a supply of substrates for the synthesis of milk components and the initiation and maintenance of all of the mechanisms of lactation in the secretory cells. The pathways for protein, lactose, and fat synthesis in the mammary cell require energy in the form of ATP used for the common relationship of the system. Many of the cellular subcomponents play a direct role in the synthesis of certain milk constituents and may be involved with more than one. The relationship of the systems to the major pathway in the mammary cell involving a complex interrelationship has been proposed [3].

Milk proteins are classified into caseins and non-casein (whey) proteins. Caseins also are milk-specific proteins. There are four basic types of molecules: the αs_1, αs_2, β, and κ in bovine milk which each type comprises many molecular species [4]. K-casein is a glycoprotein that stabilizes the micelle against precipitation by calcium ions. Non-casein (whey) proteins comprise both milk-specific and serum proteins which in some species (e.g. rodents) contain few specific proteins while carnivores contain many. Although the basic mechanism of milk protein synthesis is identical in all species. Variations in the activity of which would be expressed in different milk protein compositions. The milk-specific protein like serum protein occurs in milk, especially immunoglobulins being high in colostrum in many species (e.g. cows, sheep, and pigs) in which the structure of the placental membranes prevents their transfer from the maternal to the fetal plasma. The question then arises, whether this protein in milk comes from blood plasma or by the synthesis in plasma cells situated in the mammary gland via either transcellular route or paracellular route. The milk-specific non-casein proteins are α-lactalbumin and β-lactoglobulin. The role of α-lactalbumin is known to involve in lactose biosynthesis and increase milk yield. This example indicates that complex mechanisms of metabolic pathways occur in the mammary gland. The elucidation of these and many other metabolic differences among species in the synthesis of milk is an area with an important related application for some that should receive continued and increasing attention.

An increase in the synthesis of protein, lactose, and fat in milk is a complex mechanism that improves the efficiency of milk production. The achievement is proposed by combining genetic improvements with good management including improving the nutritional availability of the basic compounds used by the mammary gland to produce milk. The advanced study of molecular biology and discovery of DNA that manipulation of DNA transcription is a powerful means to affect phenotypic outcomes. In addition to molecular study transcription, many posttranscriptional regulations can greatly affect phenotype, among which phosphorylation of proteins constitutes a primary one.

3. Role of amino acid supply for milk protein synthesis

It is known that the rate of substrates supplying to the mammary gland is determined by the concentration of substrates in the plasma and mammary blood flow. Thus, the sustainable milk protein synthesis in the mammary gland is dependent upon its blood supply to provide amino acids at appropriate rates. The rate of milk protein production depends on the function of a number of secretory cells and their metabolic activity. However, the mechanisms can be divided into three main levels of regulation, which are the arterial flow of amino acids in the mammary gland, the amino acids extraction by the mammary gland via amino acids transport systems, and the metabolic and secretion activities of the mammary epithelial cell. A dynamic exchange of amino acids between extracellular and intracellular compartments is dependent on changes driven by environmental and physiological stimuli. There is evidence that the transport of amino acids to the lactating cell is a limiting step for milk protein synthesis which depends on the concentration in plasma and extraction rates by the mammary gland. Some amino acids, for example, methionine and lysine are identified as the most limiting amino acids for milk protein synthesis in dairy cows [5] as well as in humans and other species [6]. Methionine uptake by the bovine mammary gland occurs via the sodium-independent neutral and basic amino acids transporter system [7]. Amino acids transporters may also act as sensors of intracellular and extracellular amino acids abundances regulating directly or indirectly protein synthesis. The implications and the extent of the amino acids sensing by amino acids transporters in milk protein synthesis are currently unknown but may be a phenomenon worth investigating.

4. Regulation of protein synthesis by hormone supply and the energy utilization

It has been realized that during late pregnancy, lactogenesis occurs concurrently with mammary development, and many hormones are needed for maximal stimulation of lactogenesis. Many factors have been reported to be capable of influencing lactation persistency particularly in ruminants. The hormonal control of substrates uptake of the mammary gland and milk protein yield is complex. Milk protein synthesis is highly regulated by amino acids levels, amino acid transporters, and insulin via transcriptional and post-transcriptional routes, with the insulin-mTOR pathway playing a central role [8, 9].

The mammary gland is known to be hypersensitive to insulin during lactation and at the onset of pregnancy, with an increased sensitivity until the end of pregnancy, primarily due to an augmented kinase activity of the insulin receptor [10]. One of the primary functions of insulin in the mammary gland is the control of milk protein synthesis by inducing translation via activation of mTOR pathway and activation of signal transducer and activator of transcription (STAT5) through phosphorylation [9].

During lactation, the mammary gland becomes a metabolically active and the highest energetically demanding tissue. In ruminants, increases in energy utilization for milk protein synthesis are due to an increase in mRNA translation in the mammary gland [9]. In addition, the transformation of dietary nitrogen into milk proteins [11], with a high rate of protein turnover [12]. In the goat, the daily tissue protein synthesis can account for as much as 88% of the total protein synthesized, of which 50% of ATP generated by the lactating mammary glands is used for the synthesis of extra-mammary tissue (nonmilk) protein [13]. Insulin can also affect milk protein synthesis by increasing the uptake of amino acids, particularly the branched-chain and the essential amino acids [9, 14]. An increase in intracellular glucose can enhance protein synthesis by increasing ATP production with subsequent inhibition of 5′ AMP-activated protein kinase activity, one of the main negative regulators of mTOR [15].

The availability of amino acids is not only essential for milk protein synthesis, but they can also activate the translational machinery through mTOR pathway [9] and affect the expression of milk protein genes through nutrient/gene interactions [16]. Methionine is well known for its essential role in the initiation of mRNA translation, increasing protein expression and phosphorylation of STAT5a and mTOR and protein expression of casein [17]. In addition to methionine, other amino acids, such as tryptophan, arginine, and isoleucine, have positive effects on milk protein synthesis via phosphorylation of mTOR pathway-related proteins in bovine mammary epithelial cells [18].

The role of energy and amino acids, as well as amino acids transporters in controlling milk protein synthesis through posttranscriptional regulation by an insulin-mTOR signaling pathway, has been previously proposed as a detailed model in the regulation of milk protein synthesis [19]. A review on the translational regulation of milk protein synthesis, including a historical overview of the progress toward understanding post-transcriptional regulation, has been reported [20]. Among transcription factors important for the regulation of casein genes, signal transducer and activator of transcription 5 (STAT5), is the most important due to its role in controlling the expression of various caseins, genes, and lactalbumin [9]. Although insulin plays an important role in the activation of STAT5, other hormones, such as prolactin and glucocorticoids, can also regulate the activity of STAT5 [21]. Interestingly, the level of regulation of STAT5 through Jak/Stat signaling by hormones may vary across species, which has been reviewed previously [9].

5. Conclusion

Milk protein is one of the most important nutrients in milk. Research on the general mechanisms involved in the synthesis of milk protein provides evidence that the synthesis of milk protein is complex in the mammary gland and is regulated by multiple factors and the protein components in milk are specific to the mammary tissue. During lactogenesis, several hormones are needed for maximal stimulation of milk production. The role of insulin has been shown to participate in the regulation of milk protein synthesis with a pivotal role in perturbation with the mTOR and STAT5 signaling pathways. However, the control of milk protein synthesis is still needed to be completely understood, to open new insight into research not previously considered. For example, the elucidation of the mechanism of milk protein synthesis and other metabolic differences among species. High environmental temperatures are known to affect milk secretion at various levels of mechanisms both directly and indirectly in lactating animals in the tropic. Many technologies are required using to elucidate the complex mechanism of milk protein synthesis. The molecular study using omics technologies is a technic that has been introduced to provide the possibilities to further investigate related complex mechanisms of milk protein synthesis. These novel research approaches may contribute to revealing the mechanism of milk protein synthesis and the novel biomarkers in milk affected by some factors. However, the elucidation of these and many other metabolic differences among species in the synthesis of milk protein is an area with an important related application for some that should receive continued and increasing attention.

Author details

Narongsak Chaiyabutr[1,2]

1 Faculty of Veterinary Science, Department of Physiology, Chulalongkorn University, Bangkok, Thailand

2 The Academy of Science, The Royal Society of Thailand, Bangkok, Thailand

*Address all correspondence to: narongsak.c@chula.ac.th

References

[1] Mepham TB. Synthesis and secretion of milk proteins. In: Peaker M, editor. Comparative Aspects of Lactation. Vol. 41. Symposia of the Zoological Society of London; London, New York: Academic Press; 1977. pp. 57-75

[2] Shennan DB, Peaker M. Transport of milk constituents by the mammary gland. Physiological Reviews. 2000; **80**:925-951

[3] Smith GH, Taylor DJ. Mammary energy metabolism. In: Peaker M, editor. Comparative Aspects of Lactation. Vol. 41. Symposia of the Zoological Society of London; London, New York: Academic Press; 1977. pp. 95-111

[4] Thompson MP, Farrel HM. Genetic variants of the milk proteins. In: Larson BL, Smith VR, editors. Lactation. Vol. 3. New York, London: Academic Press; 1974. pp. 109-134

[5] Guinard J, Rulquin H. Effect of graded levels of duodenal infusions of casein on mammary uptake in lactating cows. 2. Individual amino acids. Journal of Dairy Science. 1994;77:3304-3315

[6] Davis TA, Nguyen HV, Garcia-Bravo R, Fiorotto ML, Jackson EM, Lewis DS, et al. Amino acid composition of human milk is not unique. The Journal of Nutrition. 1994;**124**:1126-1132

[7] Baumrucker CR. Amino acid transport systems in bovine mammary tissue. Journal of Dairy Science. 1985;**68**:2436-2451

[8] Osorio JS, Lohakare J, Bionaz M. Biosynthesis of milk fat, protein, and lactose: Roles of transcriptional and posttranscriptional regulation. Physiological Genomics. 2016;**48**:231-256

[9] Bionaz M, Hurley WL, Loor JJ. Milk protein synthesis in the lactating mammary gland: Insights from transcriptomics analyses. In: Hurley WL, editor. Milk Protein. London, UK: IntechOpen; 2012. pp. 285-324

[10] Carrascosa JM, Ramos P, Molero JC, Herrera E. Changes in the kinase activity of the insulin receptor account for an increased insulin sensitivity of mammary gland in late pregnancy. Endocrinology. 1998;**139**:520-526

[11] Bequette BJ, Backwell FR, Crompton LA. Current concepts of amino acid and protein metabolism in the mammary gland of the lactating ruminant. Journal of Dairy Science. 1998;**81**:2540-2559

[12] Bequette BJ, Backwell FR. Amino acid supply and metabolism by the ruminant mammary gland. The Proceedings of the Nutrition Society. 1997;**56**:593-605

[13] Hanigan MD, France J, Mabjeesh SJ, McNabb WC, Bequette BJ. High rates of mammary tissue protein turnover in lactating goats are energetically costly. The Journal of Nutrition. 2009;**139**:1118-1127

[14] Griinari JM, McGuire MA, Dwyer DA, Bauman DE, Barbano DM, House WA. The role of insulin in the regulation of milk protein synthesis in dairy cows. Journal of Dairy Science. 1997;**80**:2361-2371

[15] Sarbassov DD, Ali SM, Sabatini DM. Growing roles for the mTOR pathway. Current Opinion in Cell Biology. 2005;**17**:596-603

[16] Jousse C, Averous J, Bruhat A, Carraro V, Mordier S, Fafournoux P.

Amino acids as regulators of gene expression: Molecular mechanisms. Biochemical and Biophysical Research Communications. 2004;**313**:447-452

[17] Huang YL, Zhao F, Luo CC, Zhang X, Si Y, Sun Z, et al. SOCS3-mediated blockade reveals major contribution of JAK2/STAT5 signaling pathway to lactation and proliferation of dairy cow mammary epithelial cells in vitro. Molecules. 2013;**18**:12987-13002

[18] Appuhamy JA, Knoebel NA, Nayananjalie WA, Escobar J, Hanigan MD. Isoleucine and leucine independently regulate mTOR signaling and protein synthesis in MAC-T cells and bovine mammary tissue slices. The Journal of Nutrition. 2012;**142**:484-491

[19] Bionaz M, Loor JJ. Gene networks driving bovine mammary protein synthesis during the lactation cycle. Bioinformatics and Biology Insights. 2011;**5**:83-98

[20] Rhoads RE, Grudzien-Nogalska E. Translational regulation of milk protein synthesis at secretory activation. Journal of Mammary Gland Biology and Neoplasia. 2007;**12**:283-292

[21] Groner B. Transcription factor regulation in mammary epithelial cells. Domestic Animal Endocrinology. 2002;**23**:25-32

Section 2

Microbiological and Biochemical Characteristics of Milk Production

Chapter 2

Microbiological Quality and Biochemical Characteristics of Lactic Acid Bacteria from Camel Milk as Affected by the Production System and Stage of Lactation

Imen Fguiri, Manel Ziadi, Amel Sboui, Naziha Ayeb, Moufida Atigui, Samira Arroum, Mohamed Hammadi and Touhami Khorchani

Abstract

The aim of this work is to study the effect of lactation stage and camel farming system on microbiological, physicochemical parameters, and identification of lactic acid bacteria (LAB) of camel milk. Samples were collected from four camels in semi-intensive system and four camels in intensive system. Microbiological and physicochemical parameters were analyzed. Furthermore, to study the effect of lactation stage, samples were collected from three camels and followed during a period of 10 months of lactation from parturition. LAB were isolated from this sample and identified by biochemical methods. The difference between the physico-chemical characteristic basis of camel farming system are not statistically different except fat. The microbiological analysis showed a significant difference in total mesophilic bacteria, yeast, and molds and total coliform between intensive, semi-intensive, and extensive system. The difference between physicochemical and microbiological characteristics basis of lactation stage are statistically significant. In the intensive system, they were identified the same genre of bacteria: *Lactococcus lactis*, but in semi intensive system, we found different species of LAB. Eight of LAB identified as different *Lactococcus* or *Lactobacillus* was isolated in colostrums. The diversity of LAB was affected by lactation stage and farming system.

Keywords: camel milk, production system, lactation stage, lactic acid bacteria, milk proteins

1. Introduction

The dromedary camel (*Camelus dromedarius*) is the most important animal in the arid areas in the world. It is a multipurpose animal, used for its supply of milk, meat, hides, and transport [1].

Camel milk has a sweet and sharp taste normally, but at times it can taste salty and other times it tastes watery. The quality of milk is affected by the age of the animal, the stage of lactation, the quality and quantity of feed, as well as the amount of water available [2].

In Tunisia, camel breeding is conventionally extensive; a method perfectly suited to the biology of the specie and it is concentrated in the southern areas [3]. In addition to the extensive system; a new breeding method was developed in several places in the world that could be described as intensive system. This system is based on a set of techniques and ways to optimize production capacity of the animal [4].

Furthermore, there is the semi-intensive also called integrated system which was created due to the decrease in pasture and feed [5].

Camel milk contains inevitably microflora, the nature and significance are determined by the health status of the animal, milking conditions, temperature, etc. Even under rigorous collection conditions, the number of microflora does not exceed 5 10^3 cells/ml [6] and this can be due to the inhibition of pathogens bacteria properties in camel milk [7].

Indeed, Al-Mohizea et al. [8] concluded that the hygienic quality of camel milk is satisfactory based on counts of four groups of microorganisms (total aerobic flora, psychrotrophic, coliform, and sporulating bacteria).

The dominant and beneficial microflora in camel milk represented by lactic acid bacteria (LAB) is a potential source of biological materials to be used in dairy technology [9]. LAB strain characterized by their ability to transform lactose and to improve the digestibility of fermented dairy products [10] as well as to preserve [11]. They were also employed for improvement of the taste, texture, and viscosity in the manufacture of dairy products [12]. The ability of LAB to produce probiotics [13] and stimulation of the immune system [14] render this group of microorganisms' essential importance dairy industry which gives added values for dairy product.

The effect of lactation stage and farming system on the physicochemical composition of milk has been the subject of some works [15, 16]. However the microbiological quality is not well studied.

The dominant and beneficial microflora in camel milk are mainly LAB. This group of bacteria is considered to be a potential source of biological agents for use in dairy technology [9]. This study aimed to determine the impacts of lactation stage, production system on physicochemical characteristics and microbiological quality especially the concentration of LAB in camel milk.

2. Material and methods

2.1 Source of sampling

2.1.1 Effect of breeding system

Eight camels (*C. dromedarius*) Maghrebi breed belonging to the herd of the Arid Lands Institute (Medenine) were studied: four camels in semi-intensive system (Medenine) and four camels in intensive system (Chenchou station, Gabès). Camels were followed during the sixth and ninth months of lactation. In addition to these two types samples were collected from four camels reared in extensive system on El Ouara (Ben Ghilouf, Tataouine) and brought to the laboratory of Livestock and Wildlife for physicochemical and microbiological analyses (**Table 1**).

Rearing method	Feeding
Intensive system	5 kg hay alfalfa
	5 kg of oat hay
	2 kg concentrate No. 7 (cow milk)
	6 kg green alfalfa
Semi-intensive system	6 hours of grazing
	Daily supplementation of mixture (2.7 kg) of: barley (31%), olive pomace (54%), and wheat bran (15%)
Extensive system	Halophytic plants

Table 1.
Different types of feeding depending on farming system.

2.1.2 Effect of lactation stage

Three Maghrebi camels belonging to the herd of the Arid Lands Institute (Medenine) were monitored for a period of 12 months from parturition. The stage of lactation was subdivided in four periods:

i. First period was colostrums phase: Samples were collected daily in the first week.

ii. Second period (early lactation): Samples were collected weekly between second weeks and second month.

iii. Third period (mid-lactation): Samples were collected between once a month from the third to eighth month.

iv. Fourth period (end of lactation): Samples were collected once a month during the rest of lactation.

All the physicochemical and microbiological analysis was performed in the Laboratory of Livestock and Wildlife.

2.2 Physicochemical analyses

pH and acidity of milk were measured immediately after arrival at the laboratory. The viscosity (in cP) was determined by a Brookfield type viscometer (model DV-E, MA, USA). Dry matter, ash, and total nitrogen contents were determined by dry combustion in a furnace (550°C) that was purged with O_2 gas according to International standard methods [17]. The fat content was measured by an acid-butyrometric method using a "neusol solution."

2.3 Microbiological analysis

Total aerobic mesophilic flora was carried out on plate count agar (PCA; Scharlau Chemie S.A.), incubated at 37°C for 72 h. Yeast and molds on Sabouraud Chloramphenicol (Pronadisa) and incubated at 25°C for 3–5 days. Total coliforms were grown in violet red bile agar (AppliChem) in double layer. LAB were plated on De Man-Rogosa-Sharpe (MRS) agar (Scharlau Chemie S.A.) and incubated at 30°C for 48 h.

2.4 Isolation and identification of LAB

LAB was isolated on MRS agar (Pronadisa) and incubated at 30°C for 24–48 h in order to apply the conventional tests for identification. All isolates were initially examined for Gram staining and catalase reaction. Only Gram-positive and catalase-negative isolates were considered. The biochemical identification was carried out using API systems. API 50 CH was used in conjunction with API 50 CHL medium for the identification of *Lactobacillus* and related genera strips according to the manufacturer's instructions (Biomerieux, Marcy-l'Etoile, France) [18].

2.5 Statistical analysis

SAS software (version 9) was used for statistical analysis of data. Production system and lactation stage were analyzed by ANOVA using general linear model (GLM) for determination their effect for physicochemical and microbiological characteristics. The means values were compared using SNK test.

3. Results and discussion

3.1 Effect of production system

3.1.1 The physicochemical characteristics

The values of pH for the three production systems (i.e., intensive, semi-intensive, and extensive) averaged 6.40, which is relatively similar to the average pH value (6.43 ± 0.07) reported by [19] for Maghrebi Libyan camels kept different systems (intensive and extensive and slightly higher than that reported by [20] for raw milk (pH = 6.0). The minimum value of pH was observed in extensive system, and this is might be related to the high content of LAB in milk collected from camels under the same system (**Table 2**). The production system had no effect ($P > 0.05$) on DM and ash contents of milk. Dry

Parameter	Production system			Significance
	Intensive	Semi-intensive	Extensive	
pH	6.46 ± 0.16^{a}	6.46 ± 0.17^{a}	6.29 ± 0.088^{a}	NS
Acidity (°D)	16.75 ± 1.83^{a}	16.12 ± 2.41^{a}	17.50 ± 2.08^{a}	NS
Viscosity (cP)	3.65 ± 0.54^{a}	3.85 ± 0.87^{a}	4.62 ± 1.06^{a}	NS
Fat (g/L)	21.37 ± 6.90^{b}	26.00 ± 8.78^{a,b}	34.75 ± 10.68^{a}	*
Dry matter (g/L)	117.20 ± 10.28^{a}	118.70 ± 7.75^{a}	116.30 ± 8.27^{a}	NS
Ash (g/L)	9.56 ± 1.80^{a}	9.79 ± 1.65^{a}	9.44 ± 1.80^{a}	NS
Protein (g/L)	31.59 ± 2.48^{c}	35.86 ± 4.21^{b}	43.65 ± 4.00^{a}	**

^{a,b,c}*Means in the same line with the same letter are not statistically different ($P > 0.05$).*
NS, non-significant.$^{*}P < 0.05$.*
$^{**}P < 0.01$.

Table 2.
Effect of breeding system on physicochemical parameters of camel milk.

matter values were similar to those reported by [19] for milk representing intensive and extensive production systems (117.2 vs. 116.3 g/L), while the ash content was slightly higher (9.56 vs. 9.44 g/L). The extensive system presented a highest value of fat and protein contents. However, raising camels under intensive or extensive systems had no effect of fat content but milk from the extensive system had lower protein content (24.5 vs. 31.9 g/L) [19].

3.1.2 The microbiological characteristics

Significant differences were observed in the microbial load among the different production systems. The highest bacterial load was marked in extensive system, which can be due to the environment, processing condition and a transportation time from milking to analysis (**Table 3**). This result is similar for cow milk [21] who reported that the milk quality is affected by production system of livestock. The presence of LAB in camel milk was predictable because milk provides optimal natural environment for the growth of this group of bacteria whatever the source of milk is (sheep, goat, and cattle).

3.2 Effect of lactation stage

3.2.1 The physicochemical characteristics

It was observed that the pH value of camel milk was significantly affected during lactation (**Table 4**). Colostrum presented a low pH, due to high proteins contents [22]. Post-partum changes in gross chemical composition of camel milk showed an increase in fat. In late phase of lactation, the fat was significantly higher than in the early phase of lactation. The variation in protein content during the period of lactation was similar with result reported by [23].

3.2.2 The microbiological characteristics

The TPC fluctuated during lactation stages: increased in the early lactation stage followed by decrease in the mid-lactation before increasing again at the end, the values of TPC ranged between 2.64 and 2.30 $\log_{10}$ CFU/ml (**Table 5**). The yeast and molds content in Moroccan camel's milk was found to be higher with an

	Production system			**Significance**
	Intensive	**Semi-intensive**	**Extensive****	
TAPC ($\log_{10}$ CFU/ml)	2.96 ± 1.24^b	4.13 ± 0.98^b	5.47 ± 0.64^a	**
Yeast/mold ($\log_{10}$ CFU/ml)	1.36 ± 1.25^b	0.45 ± 0.64^b	6.36 ± 2.44^a	**
Total coliform ($\log_{10}$ CFU/ml)	2.17 ± 1.68^b	2.53 ± 1.53^b	5.38 ± 0.26^a	**
LAB ($\log_{10}$ CFU/ml)	0.13 ± 0.08^a	1.13 ± 0.20^a	1.43 ± 0.87^a	NS

^{a,b}Means within the same line with different letter are statistically different ($P < 0.05$).
*NS, non-significant. **$P < 0.01$.*

Table 3.
Effect of production system on microbiological parameters.

Parameters	Lactation stage				Significance
	First week	Early lactation	Mid-lactation	End last lactation	
pH	6.35 ± 0.18^{b,c}	6.45 ± 0.13^{a,b}	6.52 ± 0.13^{b}	6.64 ± 0.11^{a}	**
Viscosity (cP)	6.61 ± 2.55^{a}	5.82 ± 3.12^{a,b}	3.37 ± 0.45^{c}	4.17 ± 0.53^{b,c}	**
Fat content (g/L)	11.72 ± 9.78^{c}	20.30 ± 5.19^{b}	21.40 ± 4.77^{b}	28.44 ± 8.67^{a}	**
Dry matter (g/L)	127.70 ± 12.38^{a}	113.35 ± 8.16^{b}	106.09 ± 5.24^{c}	105.68 ± 4.01^{c}	**
Ash (g/L)	10.74 ± 2.42^{a}	7.95 ± 1.50^{b}	7.07 ± 1.23^{b}	6.96 ± 0.89^{c}	**
Acidity (°D)	24.78 ± 5.63^{a}	17.39 ± 4.04^{b}	15.40 ± 3.08^{b}	16.33 ± 2.64^{b}	**
Proteins (g/L)	43.07 ± 2.11^{a}	33.85 ± 2.26^{b}	27.93 ± 0.87^{d}	31.34 ± 1.61^{c}	**

^{a,b,c,d}*Means within a line with different letter are statistically different (P < 0.05).*

Table 4.
Effect of lactation stage on physicochemical characteristics.

Parameters	First week	Early lactation	Mid-lactation	Late lactation	Significance
TAPC (log UFC/ml)	2.64 ± 0.80^{a}	2.30 ± 1.05^{a}	2.48 ± 0.99^{a}	2.41 ± 0.76^{a}	NS
Yeast and molds (log UFC/ml)	2.14 ± 0.99^{a}	2.17 ± 0.30^{a}	1.65 ± 0.82^{a,b}	1.28 ± 1.24^{b}	NS
LAB (log UFC/ml)	2. 79 ± 0. 53^{a,b}	2.00 ± 1.05^{b,c}	1.62 ± 1.23^{c}	2.48 ± 0. 66^{a}	**

^{a,b,c}*Means within a line with different letter are statistically different (P < 0.05).*
*NS, non-significant. **P < 0.01.*

Table 5.
Effect of lactation stage on microbiological characteristics.

average count of 4.6 log_{10} CFU/ml [12]. LAB counts ranged between 1.62 and 2.79 log_{10} CFU/ml and the difference between lactation stages being significant with a maximum in colostrum stage. LAB was the predominant microflora in camel milk since it has been proved that they are capable of producing inhibitory substances other than organic acids (lactate and acetate) that are antagonistic toward other microorganisms [13].

3.3 Isolation and identification of LAB

Regarding the carbohydrates fermentations the strains were divided in two groups (**Table 6**). The first ones dominated by regular rods ($SCC_{1,8}$, $SCC_{1,7}$, $SCC_{1,15}$, and $SCC_{1,2}$) were tentatively identified as *Lactobacillus plantarum*, *Lactobacillus pentosus*, and *Lactobacillus brevis*. The second group was coccoid in shape (SLC_{ch14}, SLC_{ch6}, $SCC_{1,13}$, $SCC_{1,33}$, and $SCC_{1,6}$). They were tentatively identified as *Lactococcus lactis* and *Pediococcus pentosaceus*. Earlier studies have been reported the presence of the *L. plantarum* and *L. brevis* in Sudanese fermented camel milk [24]. Sun et al. [25] isolated the *L. plantarum* and *L. lactis* from traditional fermented milk in Mongolia.

Strains	SLC_{ch6}	$SCC_{1,7}$	SLC_{ch14}	$SCC_{1,13}$	$SCC_{1,33}$	$SCC_{1,15}$	$SCC_{1,24}$	$SCC_{1,6}$	$SCC_{1,8}$	$SCC_{1,2}$
Glycerol	+	−	W	W	+	W	−	W	W	+
L-Sorbose	−	−	−	−	W	−	−	−	−	−
D-Sorbitol	−	−	−	−	−	+	−	+	−	−
Amygdaline	−	+	+	W	W	+	−	+	−	−
Esculine	+	+	W	W	+	+	W	W	+	W
D-Melezitose	−	−	−	−	−	−	−	+	−	W
Amidon	−	W	+	W	+	−	W	−	W	W
Identification	*L. lactisssp lactis1*	*Lb plantarum*	*L. lactis ssp lactis1*	*L. lactis ssp lactis1*	*L. lactis ssp lactis1*	*Lb pentosus*	*L. lactis ssp lactis1*	*Pediococcus pentosaceus*	*Lb plantarum*	*Lb brevis*

+, positive; W, weakly positive; −, negative after 48 h of incubation at 37°C; SLC_{ch}, Strain Milk Camel Chenchou; SCC, Strain colostrum camel.

Table 6.
Fermentation profiles of LAB isolated for camel milk.

4. Conclusion

The present study showed variations of physicochemical and microbiological characteristics in camel milk was affected by production systems and stages of lactation. Physicochemical characteristics of camel milk samples obtained from different production system revealed highly significant variations between these systems in the content of fat and protein. Additionally, stages of lactation showed variations in the physicochemical and microbiological characteristics of camel milk. LAB were also affected by production system and lactation stage.

Acknowledgements

Authors are grateful to central laboratory technicians in Arid Lands Institute for her collaboration in biochemical analysis.

Conflict of interest

The authors declare that there is no conflict of interests regarding the publication of this paper.

Author details

Imen Fguiri[1*], Manel Ziadi[2], Amel Sboui[1], Naziha Ayeb[3], Moufida Atigui[4], Samira Arroum[1], Mohamed Hammadi[1] and Touhami Khorchani[1]

1 Laboratory of Livestock and Wild Life, Institute of Arid Lands (IRA Medenine), Medenine, Tunisia

2 Laboratory of Microbial Technology and Ecology (LETMi), National Institute of Applied Sciences and Technology (INSAT), Tunis, Tunisia

3 Regional Center for Agricultural Research (CRRA) Sidi Bouzid, Tunisia

4 Higher School of Agriculture Mateur, University of Carthage, Tunisia

*Address all correspondence to: imen.fguiri@yahoo.com

References

[1] Souilem O, Barhoumi K. Physiological particularities of dromedary (*Camelus dromedarius*) and experimental implications. Scandinavian Journal of Laboratory Animal Science. 2009;**36**:19-29

[2] Abdel Rahman IE, Dirar HA, Osman A. Microbiological and biochemical changes and sensory evaluation of camel milk fermented by selected bacterial starter cultures. African Journal of Food Science. 2009;**3**:398-405

[3] Moslah M, Megdiche F. L'élevage camelin en Tunisie. Cahiers Options Méditerranéennes Série A. 1989;**2**:33-36

[4] Faye B, Grech S, Khorchani T. Le dromadaire, entre, feralisation et intensification. Anthropozoologica. 2004;**39**(2):7-14

[5] Diallo BC. L'élevage du dromadaire en Mauritanie. Options Méditerranéennes—Série. Séminaires—n°2. Dans: Actes du colloque de Rabat-Maroc; 1989. pp. 29-32 [Accessed: 25-27 Octobre 1988]

[6] Larpent JP, Copin MP, Germonville A, Jacquet M, Thetas JL. Microbiologie du lait et des produits laitiers. In: Larpent LP, editor. « Microbiologie alimentaire ». Tec. Doc. 1ère ed. Paris: Lavoisier; 1997

[7] Barbour K, Nabbut NH, Frerichs WN, Al Nakhli HM. Inhibition of pathogenic bacteria by camel's milk: Relation to whey lysozyme and stage of lactation. Journal of Food Protection. 1984;**47**:838-840

[8] Al-Mohizea IS, Mashhadi AS, Fawwal A, Al-Shalhat A. Microbiological and shelf life assessment of chilled eviscerated whole chicken broilers in Saudi Arabia. British Poultry Science. 1994;**35**:519-526

[9] Khedid K, Faid M, Mokhtari A, Soulaymani A, Zinedine A. Characterization of lactic acid bacteria isolated from the one humped camel milk produced in Morocco. Microbiological Research. 2009;**164**:81-91

[10] Weinberg Z, Shatz O, Chen Y, Yosef E, Nikbahat M, Ben-Ghedalia D, et al. Effect of lactic acid bacteria inoculants on in vitro digestibility of wheat and corn silages. Journal of Dairy Science. 2007;**90**:4754-4762

[11] Abdelbasset M, Djamila K. Antimicrobial activity of autochthonous lactic acid bacteria isolated from Algerian traditional fermented milk "Raïb". African Journal of Biotechnology. 2008;7(16):2908-2914

[12] Soukoulis C, Panagiotidis P, Koureli R, Tzia C. Industrial yogurt manufacture: Monitoring of fermentation process and improvement of final product quality. Journal of Dairy Science. 2007;**90**:2641-2654

[13] Temmerman R, Pot B, Huys G, Swings J. Identification and antibiotic susceptibility of bacterial isolates from probiotic products. International Journal of Food Microbiology. 2002;**81**:1-10

[14] Kalliomäki M, Salminen S, Arvilommi H, Kero P, Koskinen P, Isolauri E. Probiotics in primary prevention of atopic disease: A randomised placebo-controlled trial. The Lancet. 2001;**357**:1076-1079

[15] El-Hatmi H, Khorchani T, Attia H. Characterization and composition of camel's (*Camelus dromedarius*) colostrum and milk. Microbiology. 2006;**18**:13-17

[16] Shuiep ES, El Zubeir IEM, El Owni OAO, Musa HH. Influence of

season and management on composition of raw camel (*Camelus dromedarius*) milk in Khartoum state, Sudan. Tropical and Subtropical Agroecosystems. 2008;**8**:101-106

[17] AFNOR. Contrôle de la qualité des produits alimentaires: lait et produits laitiers: analyses physicochimiques. Paris La Défense; 1993. 581 p

[18] Ghanbari M, Rezaei M, Jami M. Isolation and characterization of *Lactobacillus* species from intestinal contents of beluga (Husohuso) and Persian sturgeon (Acipenserpersicus). Iranian Journal of Veterinary Research. 2009;**10**:152-157

[19] Alwan OA, Igwegbe AO, Ahmad AA. Effects of rearing conditions on the proximate composition of Libyan Maghrebi camels' (*Camelus dromedarius*) milk. International Journal of Engineering and Applied Sciences. 2014;**4**(8):1-6

[20] Ghouri M, Afshan N, Javed S, Aziz F, Sadat A, Chohan A, et al. Physiochemical evaluation and liability of dromedary camel's milk in combating various pathogens. African Journal of Microbiology Research. 2016;**10**(41):1739-1745

[21] Sraïri MT, Hasni Alaoui I, Hamama A, Faye B. Relations entre pratiques d'élevage et qualité globale du lait de vache en étables suburbaines au Maroc. Revue de Médecine Vétérinaire. 2005;**156**(3):155-162

[22] Benkerroum N, Boughdadi A, Bennani N. Microbiological quality assessment of Moroccan camel's milk and identification of predominating lactic acid bacteria. World Journal of Microbiology and Biotechnology. 2003;**19**:645-648

[23] Aljumaah RS, Almutairi FF, Ismail E, Alshaikh MA, Sami A, Ayad M. Effect of production system, breed, parity, and stage of lactation on milk composition of dromedary camels in Saudi Arabia. Journal of Animal and Veterinary Advances. 2012;**11**(1):141-147

[24] Ashmaig A, Hasan A, El Gaali E. Identification of lactic acid bacteria isolated from traditional Sudanese fermented camel's milk (Gariss). African Journal of Microbiology Research. 2009;**3**:451-457

[25] Sun ZH, Liu WJ, Zhang JC, Yu J, Gao W, Jiri M, et al. Identification and characterization of the dominant lactic acid bacteria isolated from traditional fermented milk in Mongolia. Folia Microbiologica. 2010;**55**:270-276

Section 3

Cellular and Molecular Milk Protein Synthesis

Chapter 3

Superior Stimulation of β-Casein mRNA Accumulation by Pseudophosphorylated Prolactin: Enhanced Transcription and Message Stabilization

Wei Wu, Changhui Deng, Jennifer L. Brockman, Linda A. Schuler and Ameae M. Walker

Abstract

A proportion of secreted pituitary prolactin (PRL) is phosphorylated. However, because most commercial sources of PRL are recombinant proteins without post-translational modification, the importance of PRL phosphorylation to the production of milk proteins is an understudied area. Here, we have examined the effect of PRL phosphorylation on expression of the milk protein, β-casein, using a phospho-stable mimic of the phosphorylated form (S179D-PRL) and analyzing promoter activation and mRNA stability over a 7-day treatment period in response to this and unmodified PRL. At equivalent concentrations, the phospho-mimic showed a superior ability to activate a −2300 → +490 region of the promoter, but not an artificial promoter consisting of three Stat5 consensus sites upstream of a minimal promoter. Unlike unmodified PRL, S179D-PRL was also able to stabilize β-casein mRNA. These effects of S179D-PRL were eliminated by incubation in the MAP kinase pathway inhibitor, U0126, bringing promoter activation down to the level seen with unmodified PRL and essentially eliminating the effect on mRNA stability. These results support an important role for the posttranslational phosphorylation of PRL and signaling through the MAP kinase pathway in the production of this milk protein.

Keywords: hormone regulation, posttranslational modification, phosphorylated prolactin, molecular mimicry, S179D prolactin, beta-casein, mRNA stabilization, length of promoter

1. Introduction

PRL phosphorylation in the pituitary is regulated physiologically [1, 2]. Phospho-PRL has been demonstrated in rat [3], mouse [4], sheep, avian [5], bovine [6], and human [7] pituitary extracts. The phosphorylated form is very stable and cleared from the circulation with similar kinetics to the unmodified hormone [6–8]. Differential function analysis of unmodified and phospho-PRL demonstrated different biological activities [9, 10]. Recent studies have utilized a molecular mimic of phospho-PRL to prevent conversion to the unmodified form during the course of an experiment. This mimic was made by substituting an aspartate residue for the normally phosphorylated serine [11, 12], thereby producing S179D-PRL.

Previous work from our laboratories has demonstrated different activities for unmodified PRL (U-PRL) and S179D-PRL in the control of proliferation in the mammary gland and mammary cells in culture, with U-PRL promoting cell proliferation and S179D-PRL antagonizing this effect [12–16]. Despite these demonstrations, which indicate that mammary cells recognize and respond to the two kinds of PRL differently, there has since been very limited investigation of the importance of PRL phosphorylation to the production of milk proteins. In large part, this is likely due to the lack of availability of purified phosphorylated PRL, another reason we developed the phospho-mimic, S179D-PRL.

Using S179D-PRL, we have previously shown [14] that compared to U-PRL this phospho-mimic had a superior ability to stimulate β-casein expression. This was surprising since S179D-PRL has an inferior ability to activate Stat5 [13], activation of which is crucial to milk protein production [17].

Here, we have investigated the molecular mechanisms resulting in superior β-casein expression in response to S179D-PRL by examining the relative roles of transcriptional and posttranscriptional activities and MAP kinase signaling to the accumulation of β-casein mRNA in response to each PRL.

2. Materials and methods

2.1 Mammary cell culture and differentiation

HC11 cells were a gift from Nancy Hynes (Friedrich Miescher Institute, Basel, Switzerland). Cells were grown in RPMI 1640 (Invitrogen, Carlsbad, CA, USA) containing 10% FBS, 5 μg/ml insulin (Sigma, St. Louis, MO, USA), 10 ng/ml epidermal growth factor (Invitrogen), 100 U/ml penicillin, and 100 μg/ml streptomycin. Once confluent, they were grown 3 more days with daily medium changes. The medium was then removed, and the cells were washed five times with RPMI 1640. Cells were then treated with priming medium-RPMI 1640 supplemented with 10% charcoal-stripped horse serum (Cocalico Biologicals, Reamstown, PA, USA), antibiotics, 10 μg/ml insulin, 1 μg/ml hydrocortisone (Sigma) for 24 h followed by induction in priming medium plus U-PRL, or S179D-PRL (changed daily). This essentially follows Taverna et al. [18] except that 1 μg/ml, which maximally stimulates the endpoints measured here, instead of 5 μg/ml PRL, was used in this study. Potential differences in the uptake or degradation of U-PRL and S179D-PRL were monitored by ELISA and Western blot of media samples.

2.2 Transfection and β-casein luciferase assays

Primed HC11 cells were transfected with −2300 → +490 of the proximal rat β-casein promoter [19] subcloned into pGL3Basic (Promega, Madison, WI, USA) and a CMV-β-galactosidase construct. Sub-confluent cultures in six-well plates were transfected with β-casein luciferase DNA (2 μg), β-gal DNA (0.5 μg), and 10 μl lipofectamine/5 ml. After transfection, cells were treated with one or other form of PRL for 24 h in the absence or presence of U0126 (10 μM). For the 7-day experiment, cells were treated with the PRLs for 6 days, with daily medium changes, prior to transfection. For some experiments, cultures were transfected with an artificial promoter consisting of three Stat5 consensus sites upstream of a minimal promoter [20].

2.3 β-Casein mRNA stability

HC11 cells were induced for 1–7 days with daily medium changes and RNA was isolated. To test the mRNA stability, the transcription inhibitor 5,6-dichloro-1-b-D-ribofuranosylbenzimidazole (DRB) (Promega) was added (50 μg/ml) in the continued presence of the PRLs and in the absence and presence of the MEK1/2 inhibitor, U0126 (10 μM). Real-time PCR reactions were 12.5 μl SYBR Green PCR master mix (Applied Biosystems), 2.5 μl of 10 μM mouse β-casein forward primer (5′-CCC GTC CCA CAA AAC ATC CAG CC-3), and 2.5 μl reverse primer (5′-ATT AGC AAG ACT GGC AAG GCT G-3′), or 2.5 μl of mouse GAPDH forward primer (5′-CCA TGG AGA AGG CTG GGG-3′), and 2.5 μl reverse primer (5′-CAA AGT TGT CAT GGA TGA CC-3′), 1 μl diluted RT product and 6.5 μl ddH_2O with 10 min 95°C followed by cycles of 95°C, 1 min; 55°C, 30 s; 72°C, 45 s; and 80°C, 10 s. Both annealing temperature and Tm of mouse β-casein and GAPDH primers are similar, allowing co-amplification and a comparative C_T method for quantification of gene expression calculated by $2^{-\Delta\Delta C_T}$.

2.4 Statistical analysis

Data were subjected to ANOVA with posttests comparing specific groups and Bonferroni corrections for multiple comparisons. Data are presented as mean ± SEM. The minimal number of experiments and replicates within each experiment was 3. Analysis of the real-time RT-PCR data was as per ABI PRISM 7700 Sequence Detection System User Bulletin #2.

3. Results

Real-time RT-PCR allowed us to determine steady-state transcript levels as a function of time in response to U-PRL and S179D-PRL (**Figure 1**).

β-Casein mRNA increased over the 7-day period in response to each PRL. S179D-PRL was more efficacious than U-PRL, resulting in over twice the level of β-casein transcripts after 3 days, and 3–4 times higher levels after the full 7-day incubation. In order to determine whether these were effects on promoter activity, we utilized a β-casein promoter-luciferase construct. Importantly, this construct included the −2300 to +490 region of the β-casein promoter and not the usually employed, much smaller −344/−1 portion. As can be seen in **Figure 2A**, both PRLs stimulated reporter activity after 1 day of exposure. However, S179D-PRL was twice as effective as U-PRL. This was not the result of differential stability of these PRL forms since examination of the 24-h media from these incubations by ELISA and Western blot using an antibody, which recognizes both forms equally showed no evidence of different uptake or degradation of the PRLs (data not shown). Conduct of this experiment in the presence of U0126 demonstrated that the ERK pathway was important for the superior ability of S179D-PRL to activate the promoter (**Figure 2A**). In contrast, this inhibitor had no significant effect on U-PRL-stimulated activity and did not alter controls.

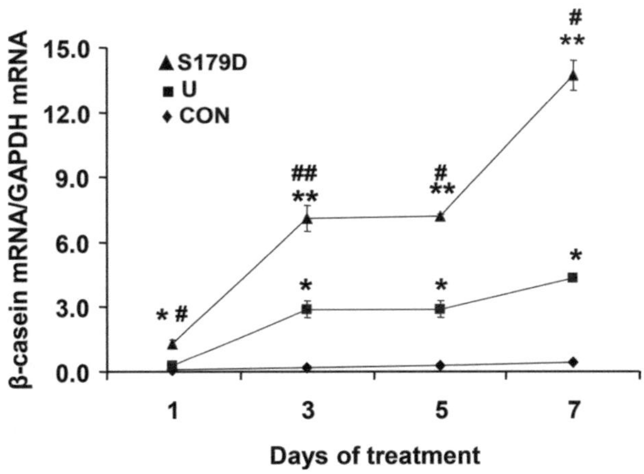

Figure 1.
*Effect of U-PRL and S179D-PRL on β-casein mRNA levels as a function of days of stimulation. *, $p < 0.01$ versus control (CON); #, $p < 0.01$ for S179D-PRL (S179D) versus U-PRL (U); ** $p < 0.001$ versus control; ##, $p < 0.05$ for S179D-PRL versus U-PRL.*

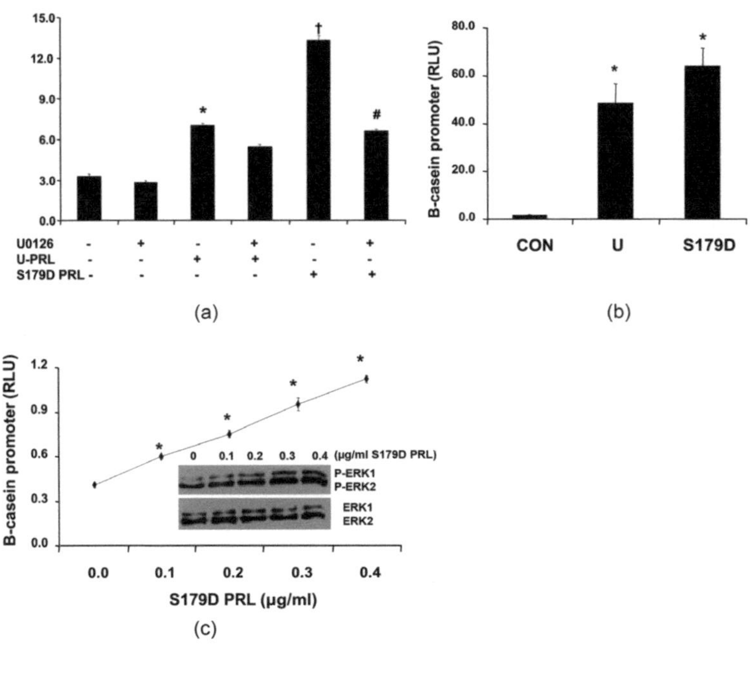

Figure 2.
*Effect of U-PRL and S179D-PRL on the −2300 → +490 β-casein promoter-luciferase. A, transfection at day 0 and assay at day 1 of treatment with and without inhibition by U0126; B, transfection at day 6 and assay at day 7 of treatment; C, ERK activation and β-casein-luciferase activity in response to S179D-PRL. *, $p < 0.01$ for U-PRL and †, $p < 0.001$ versus control. #, $p < 0.01$ versus S179D-PRL.*

The relationship between both ERK activation and promoter activity and dose of S179D-PRL is shown in **Figure 2C**.

To determine the importance of Stat5 in the differential effects of the two PRLs at the promoter, we examined their relative activities at an isolated Stat5 enhancer.

Although both PRLs were able to significantly increase activity, S179D-PRL was slightly less effective (**Figure 3**). This result demonstrated that other regulatory

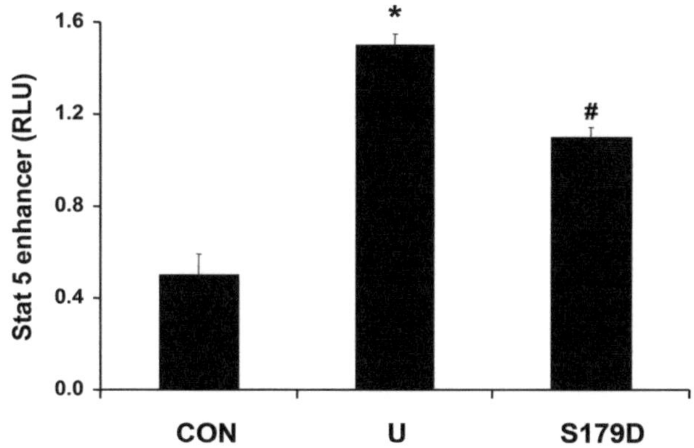

Figure 3.
*Effect of U-PRL and S179D-PRL on day 1 activity of the Stat5 enhancer. *, $p < 0.01$ and #, $p < 0.05$ versus control.*

sequences and mediators, in addition to Stat5, were important in the S179D-PRL response in the physiologic context of the β-casein promoter.

By 7 days of incubation, the difference in efficacies of the two PRLs in stimulating β-casein promoter activity was essentially eliminated (**Figure 2B**) and the degree of promoter stimulation in response to both ligands was increased. Based on the results in **Figure 1**, one would not have predicted the loss of the difference between U-PRL

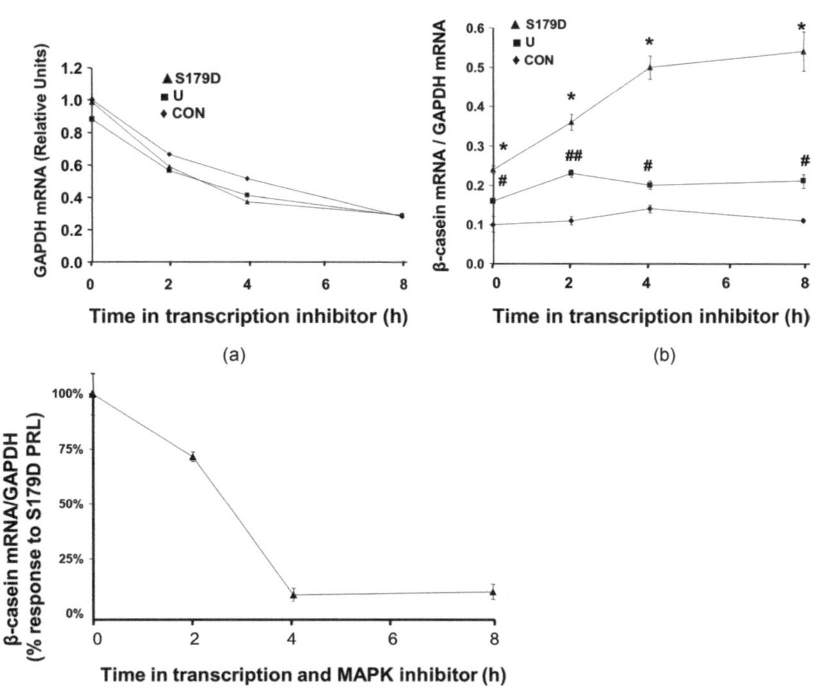

Figure 4.
*Effect of U-PRL and S179D-PRL on mRNA half-life. Following a 6-day incubation in each PRL, the transcription inhibitor, DRB, was added. A, GAPDH mRNA; B and C, β-casein mRNA levels normalized to GAPDH as a function of time after addition of DRB in the absence (B) and presence (C) of U0126 (added at time 0). The ratio at time 0 in the control cells was set at 0.1 in B. *, $p < 0.001$ for S179D-PRL versus control or U-PRL; #, $p < 0.05$ for U-PRL versus control; ##, $p < 0.001$ versus control.*

and S179D-PRL at day 7 if transcription was the most important regulator of steady-state mRNA levels. We therefore analyzed β-casein mRNA stability in response to both PRLs. Cells were incubated without or with the different PRLs for 6 days, and then, the transcription inhibitor, DRB, was added. **Figure 4A**, which plots GADPH mRNA as a function of time after inhibition of transcription, shows the expected decline in mRNA levels, with a half-life of about 4 h. This was not appreciably altered by either PRL, allowing us to use GAPDH to normalize the data for RT-PCR efficiency.

As seen in **Figure 4B**, β-casein transcripts had a half-life of about 4 h since when normalized to GAPDH, the ratio was unaltered by incubation in DRB in the controls or cells incubated in U-PRL. In contrast, incubation in DRB revealed a dramatic effect of S179D-PRL on β-casein mRNA stability. Since time 0 on this graph is after 6 days of U-PRL or S179D-PRL treatment, the effects of the two PRLs on overall mRNA levels are evident prior to DRB treatment. When U0126 was added along with DRB at 0 h, the elevated levels of β-casein mRNA in response to S179D-PRL were reduced by 2 and 4 h, indicating an important role for ERKs in S179D-PRL-induced mRNA stability (**Figure 4C**).

4. Discussion

The results demonstrate that both U-PRL and S179D-PRL increase steady-state β-casein mRNA levels. However, S179D-PRL was more effective than U-PRL after shorter exposures and also elicited a later rise not found with U-PRL. This biphasic response to S179D-PRL suggested the possibility of different mechanisms of mRNA accumulation. A large body of literature has demonstrated rapid effects of PRL treatment on the β-casein promoter [21–23]. However, a second, much smaller number of papers, which have largely been forgotten in recent years, demonstrate that PRL increases β-casein mRNA stability and that this is quantitatively much more important in terms of steady-state mRNA levels than promotion of transcription [24–27]. However, these studies used pituitary extract preparations of PRL, which contained a mixture of both unmodified and phosphorylated PRL, and so the importance of PRL phosphorylation to these activities was unknown.

Regulatory sequences within the β-casein promoter are found over a fairly large region, but most investigators have limited their examination to the activity of the most proximal 344 nucleotides in reporter gene assays. However, this promoter may not detect all the responses to S179D-PRL. CREB, ATF1, and YY1 sites outside of this region can potentially be activated by ERK1/2 (reviewed in [28]). This signaling pathway is a major mediator of S179D-PRL, although U-PRL can also activate these kinases in HC11 cells to some extent [13]. In addition, increased β-casein gene expression can be achieved by removal of YY1 from the promoter [28]. We therefore utilized a −2300 → +490 fragment of the promoter, which includes the CREB, ATF1, and YY1 sites (in addition to the STAT5 site in the −344/−1 region) in an attempt to make the reporter assay more physiologically relevant. It should be noted, however, that even this larger piece does not constitute the whole promoter.

While both PRLs increased activity of this promoter, S179D-PRL was about twice as effective as U-PRL during the first-phase response. This was unexpected, since S179D-PRL is weaker than U-PRL in stimulating STAT5 tyrosine phosphorylation [13, 15], generally thought to be the most important regulator of β-casein expression. The MEK1/2 inhibitor, U0126, eliminated the difference between U-PRL and S179D-PRL on promoter activity, indicating that ERKs 1/2 are important for the superior stimulation with S179D-PRL, but not for the activity of U-PRL. With the construct containing only the Stat5 enhancer, there was no superior effect of S179D-PRL. Together, these results demonstrate the importance of other transcription factors in S179D-PRL signals to the β-casein promoter.

Increasing β-casein transcripts with duration of S179D-PRL exposure shown in **Figure 1** suggested increased promoter activation over time in culture. However, we have shown that the difference in β-casein promoter activation between S179D-PRL and U-PRL was reduced after 7 days, consistent with a mechanism other than promoter activity for the second phase of the response to S179D-PRL. Our data indicate that S179D-PRL markedly increased β-casein mRNA stability at this time, while U-PRL was without effect. Other investigators have used pituitary-derived PRL, which is a mixture of U-PRL and phosphorylated PRL [27] and hence have not been able to make this distinction.

There are a variety of mechanisms by which β-casein mRNA stability may be regulated, including the length of the poly A tail [29] and effects at the 5′ untranslated region (5′UTR) and other areas of the 3′UTR [26, 27]. ERK1/2 activation has previously been shown to be important for some proportion of the β-casein response to mixed PRL [23]. This is the first demonstration of effects on both transcriptional and posttranscriptional regulation.

Combining these results with previous work on the role of these two PRLs in cell proliferation, we suggest that optimal alveolar development and subsequent lactation are the result of well-orchestrated exposures to combinations of U-PRL, phospho-PRL, and other physiological stimuli including milk removal. U-PRL (or a lactogen acting like U-PRL) stimulates alveolar development [14] and some β-casein expression. Phospho-PRL, which has the capacity to inhibit alveolar development [14, 16], may slow alveolar development after parturition but robustly stimulate β-casein expression by virtue of its ability to increase both promoter activity and mRNA stabilization.

5. Conclusions

These results support an important role for the posttranslational phosphorylation of PRL and signaling through the MAP kinase pathway in the production of the milk protein, β-casein. Furthermore, the results are an important reminder that use of bacterially derived recombinant prolactin that does not have normal secretory pathway posttranslational modifications may not fully represent normal physiology.

Acknowledgements

This work was supported by CBCRP grant 10PB-0127 (AMW) and by NIH grants DK 61005 (AMW) and CA 78312 (LAS). The authors would like to thank Dr. Jeffrey Rosen (Baylor College of Medicine, Houston, TX) for provision of the -2300 → +490 promoter region of the β-casein gene.

Conflict of interest

The authors declare no conflict of interest that would by any measure affect their impartiality in the presentation of results.

Author details

Wei Wu[1], Changhui Deng[2], Jennifer L. Brockman[3], Linda A. Schuler[3] and Ameae M. Walker[1*]

1 Division of Biomedical Sciences, University of California, Riverside, CA, USA

2 Department of Biochemistry, University of California, Riverside, CA, USA

3 Department of Comparative Biosciences, University of Wisconsin, Madison, WI, USA

*Address all correspondence to: ameae.walker@ucr.edu

References

[1] Ho TWC, Kawaminami M, Walker AM. Secretion of phosphorylated and non-phosphorylated monomer prolactin isoforms during rat pregnancy and pseudopregnancy. Endocrine Journal. 1993;**1**:435-439

[2] Ho TWC, Leong FS, Olaso CH, Walker AM. Secretion of specific non-phosphorylated and phosphorylated rat prolactin isoforms at specific stages of the estrous cycle. Neuroendocrinology. 1993;**58**:160-165. DOI: 10.1159/000126528

[3] Oetting WS, Tuazon PT, Traugh JA, Walker AM. Phosphorylation of prolactin. Journal of Biological Chemistry. 1986;**261**(4):1649-1652

[4] Morohoshi K, Komatani Y, Harigaya T. Estrogen induces phosphorylation of prolactin through p21-activated kinase 2 activation in the mouse pituitary gland. Journal of Reproduction and Development. 2020;**66**(6):571-578. DOI: 10.1262/jrd.2020-080

[5] Aramburo C, Montiel JL, Proudman JA, Berghman LR, Scanes CG. Phosphorylation of prolactin and growth hormone. Journal of Molecular Endocrinology. 1992;**8**:183-191. DOI: 10.1677/jme.0.0080183

[6] Brooks CL, Saiduddin S. Phosphorylation of bovine prolactin eliminates luteotropic activity in the rat. Life Sciences. 1998;**63**:1281-1287. DOI: 10.1016/s0024-3205(98)00389-0

[7] Tuazon PT, Lorenson MY, Walker AM, Traugh JA. p21-activated protein kinase γ-PAK in pituitary secretory granules phosphorylates prolactin. FEBS Letters. 2002;**515**:84-88. DOI: 10.1016/s0014-5793(02)02444-4

[8] Coss D, Yang L, Kuo CB, Xu X, Luben RA, Walker AM. Effects of prolactin on osteoblast alkaline phosphatase and bone formation in the developing rat. American Journal of Physiology. 2000;**279**:E1216-E1215. DOI: 10.1152/ajpendo.2000.279.6.E1216

[9] Wang YF, Walker AM. Dephosphorylation of standard prolactin produces a more biologically active molecule: Evidence for antagonism between nonphosphorylated and phosphorylated prolactin in the stimulation of Nb2 cell proliferation. Endocrinology. 1993;**133**:2156-2160. DOI: 10.1210/endo.133.5.8404666

[10] Krown KA, Wang YF, Ho TW, Kelly PA, Walker AM. Prolactin isoform 2 as an autocrine growth factor for GH3 cells. Endocrinology. 1992;**131**:595-602. DOI: 10.1210/endo.131.2.1639009

[11] Wang YF, Liu JW, Mamidi M, Walker AM. Identification of the major site of rat prolactin phosphorylation as serine 177. Journal of Biological Chemistry. 1996;**271**:2462-2469. DOI: 10.1074/jbc.271.5.2462

[12] Chen TJ, Kuo CB, Tsai KF, Liu JW, Chen DY, Walker AM. Development of recombinant human prolactin receptor antagonists by molecular mimicry of the phosphorylated hormone. Endocrinology. 1998;**139**:609-616. DOI: 10.1210/endo.139.2.5758

[13] Wu W, Coss D, Lorenson MY, Kuo CB, Xu X, Walker AM. Different biological effects of unmodified prolactin and a molecular mimic of phosphorylated prolactin involve different signaling pathways. Biochemistry. 2003;**42**:7561-7570. DOI: 10.1021/bi034217s

[14] Kuo CB, Wu W, Xu X, Yang L, Chen C, Coss D, et al. Pseudophosphorylated prolactin (S179D PRL) inhibits growth and promotes beta-casein gene expression in the rat mammary gland. Cell and Tissue Research. 2002;**309**:429-437. DOI: 10.1007/s00441-002-0598-8

[15] Schroeder MD, Brockman JL, Walker AM, Schuler LA. Inhibition of prolactin (PRL)-induced proliferative signals in breast cancer cells by a molecular mimic of phosphorylated PRL, S179D-PRL. Endocrinology. 2003;**144**:5300-5307. DOI: 10.1210/en.2003-0826

[16] Naylor MJ, Oakes SR, Gardiner-Garden M, Harris J, Blazek K, Ho TWC, et al. Transcriptional changes underlying the secretory activation phase of mammary gland development. Molecular Endocrinology. 2005;**19**:1868-1883. DOI: 10.1210/me.2004-0254

[17] Hennighausen L, Robinson GW, Wagner KU, Liu X. Developing a mammary gland is a stat affair. Journal of Mammary Gland Biology and Neoplasia. 1997;**2**(4):365-372. DOI: 10.1023/a:1026347313096

[18] Taverna D, Groner B, Hynes NE. Epidermal growth factor receptor, platelet-derived growth factor receptor, and c-erbB-2 receptor activation all promote growth but have distinctive effects upon mouse mammary epithelial cell differentiation. Cell Growth & Differentiation. 1991;2:145-154

[19] Lee CS, Kim K, Yu DY, Lee KK. Pretreatment with glucocorticoid is essential for lactogenic induction of the bovine beta-casein/CAT expression in HC11 cells. Endocrine Research. 1998;**24**:65-77. DOI: 10.3109/07435809809031869

[20] Schroeder MD, Symowicz J, Schuler LA. PRL modulates cell cycle regulators in mammary tumor epithelial cells. Molecular Endocrinology. 2002;**16**:45-57. DOI: 10.1210/mend.16.1.0762

[21] Perrot-Applanat M, Gualillo O, Pezet A, Vincent V, Edery M, Kelly PA. Dominant negative and cooperative effects of mutant forms of prolactin receptor. Molecular Endocrinology. 1997;**11**:1020-1032. DOI: 10.1210/mend.11.8.9954

[22] Fritsche M, Mundt M, Merkle C, Jahne R, Groner B. p53 suppresses cytokine induced, Stat5 mediated activation of transcription. Molecular and Cellular Endocrinology. 1998;**143**:143-154. DOI: 10.1016/s0303-7207(98)00140-3

[23] Gao J, Horseman ND. Prolactin-independent modulation of the beta-casein response element by Erk2 MAP kinase. Cellular Signalling. 1999;**11**:205-210. DOI: 10.1016/s0898-6568(98)00067-9

[24] Chomczynski P, Qasba P, Topper YJ. Transcriptional and post-transcriptional roles of glucocorticoid in the expression of the rat 25,000 molecular weight casein gene. Biochemical and Biophysical Research Communications. 1986;**134**:812-818. DOI: 10.1016/s0006-291x(86)80493-4

[25] Houdebine LM, Devinoy E, Delouis C. Stabilization of casein mRNA by prolactin and glucocorticoids. Biochimie. 1978;**60**:57-63. DOI: 10.1016/s0300-9084(78)80198-9

[26] Poyet P, Henning SJ, Rosen JM. Hormone-dependent beta-casein mRNA stabilization requires ongoing protein synthesis. Molecular Endocrinology. 1989;**3**:1961-1968. DOI: 10.1210/mend-3-12-1961

[27] Guyette WA, Matusik RJ, Rosen JM. Prolactin-mediated transcriptional and

post-transcriptional control of casein gene expression. Cell. 1979;**17**:1013-1023. DOI: 10.1016/0092-8674(79)90340-4

[28] Shemanko CS, Groner B. Transcription factors, cofactors and target genes mediating prolactin signals. In: Horseman ND, editor. Prolactin. Norwell, MA: Kluwer Academic Press; 2001. pp. 381-404. DOI: 10.1007/978-1-4615-1683-5

[29] Kuraishi T, Sun Y, Aoki F, Imakawa K, Sakai S. The poly(A) tail length of casein mRNA in the lactating mammary gland changes depending upon the accumulation and removal of milk. Biochemical Journal. 2000;**347**: 579-583. DOI: 10.1042/0264-6021: 3470579

Chapter 4

Omics, the New Technological Approaches to the Milk Protein Researches

Zitai Guo, Lu Ma and Dengpan Bu

Abstract

With the development of technological approaches, the perturbations of biological information in gene, mRNA, proteins, and metabolites have been gathered to broaden the cognition of synthesis processes during lactation. While omics, the series of application including genomics, transcriptomics, proteomics, and metabolomics, are mostly preferred and conducted in the investigation of lactation especially the milk protein. These new technological approaches provide a complete view of the molecular regulation pathways and make it possible to systematically investigate the lactation. The aim of this chapter is to comprehensively review the advances in knowledge regarding the great progress in milk protein synthesis as well as lactation physiology and pathology mainly in dairy cows obtained from omics technologies, meanwhile the milk proteins as well as their attributes are illustrated.

Keywords: milk protein, omics technologies, dairy cows, protein synthesis

1. Introduction

Milk protein is one of the most important nutrients in milk. It contains a variety of essential amino acids required by body maintaining and is believed to have a variety of potential biological functions [1, 2]. In past few decades, plenty of research studies were conducted for improving the milk quality especially the milk protein production, while the morphological as well as physiological focused on mammary gland has been widely investigate [3–6]. However, since the synthesis of milk protein during lactation is a complex biological activity, only little perturbations in lactation can result in a certain difference in the composition and concentration of milk protein [7]. Furthermore, the interactions between proteins, genes, and factors are diversity and dynamics when performing physiological functions, which means the tradition approaches may no longer meet the requirement of current research studies. In recent years, new approaches including Omics in lactation-related research studies received extensive attention. While Omics was one of the most used approaches since it make the study possible to explain the synthesis comprehensively and systematically at the levels of DNA, RNA, proteins, and metabolites, which can assist the further indentation of factors as well as processes regulating lactation [8, 9]. To present the use of omics approach in lactation especially milk protein synthesis research in dairy cows, the relevant introductions of omics and milk protein following the great progress in lactation physiology as well as pathology by these new approaches are illustrated in the current chapter.

2. Milk protein and its attributes

Suppose milk was considered an important nutrient source due to its perspective of molecular composition, milk proteins should be the most important source of bioactive peptides. Milk proteins were proved to have higher digestibility and more suitable content of amino acids for human being [10], which contribute to more than 3% of content in milk [11]. Based on the properties of structure, function, and solubility, milk proteins are normally divided into three categories including caseins, whey proteins, and milk fat globular membrane proteins, while these complex components make the milk proteins to vary widely [12, 13]. In short, caseins account for about 80% of total milk proteins and were mainly classified by four basic types of molecules: the αs1-, αs2-, β-, and κ-caseins [14, 15]. Whey proteins are the major component of milk whey, their components mainly include β-lactoglobulin (β-LG), α-lactalbumin (α-LA), serum albumin (SA), immunoglobulin G (IgG), and also include several small molecule proteins especially low-abundance proteins, such as enzymes and metal-binding proteins [16]. Milk fat globular membrane proteins only contribute to 1–4% of milk proteins, but play the important roles in the defense mechanisms and process of cell growth in newborns [17]. So far, plenty of research studies have investigated the types, structures as well as synthesis of those components in milk proteins, which would be reviewed in later paragraphs.

2.1 Caseins

Casein is a kind of phosphorous-containing protein [18]. The serine hydroxyl group inside casein forms an ester bond to the phosphate group, which gives it the common feature of amphiphilic. Caseins classified by different types may contain different amount of phosphate groups [19]. However, all kinds of caseins keep at least one phosphate with ester bonded. Since the caseins contain amount of phosphorylated serine groups, which bind in the form of covalent bonds [20], these groups keep the form of clusters on the surface of molecules and provide conditions for its binding to calcium ions [21], which was considered to be the most important nutritional function of casein.

The αs1-, αs2-, β-, and κ-caseins are four basic types of casein molecules as mentioned above, which are also the types synthesized by the mammary gland of dairy cows. Except for major installations, there are numerous minor fractions of caseins such as γ-casein. While none of those were found in the composition of milk [22]. Though the amino acid sequence, number, and total charge of casein are not completely the same among different variant individuals, the composition of amino acids still has the following features [23–25]:

1. The amino acids with polar and nonpolar are unevenly distributed, forming obvious hydrophilic and hydrophobic regions.

2. The content of nonpolar amino acids is higher than those with polar, while the amounts of proline and lysine are higher as well.

3. The content of sulfur-containing amino acids is lower than normal.

Though the synthesis of milk protein is closely related to animal science, the research on its structures was mostly investigated in field of food science. For caseins, the research on those monomers has become the hotspot in recent years. Previous study has predicted the structure of αs-casein by molecular simulation, and its quaternary structures of caseins were investigated by Fourier transform infrared spectroscopy as well [26]. Specifically, αs1-casein has the ability of self- aggregation, but is affected by factors including condition pH as well as the ionic strength [27]. The increase of pH value will weaken its self-aggregation, to the opposite, the increase of ionic strength will enhance self-aggregation [28]. The ability of self-aggregation in αs2-casein is similar to that in αs1-casein, which mostly depends on the ionic strength, but the aggregation begins to be weakened once the strength arrives at 0.2 mol/L [29].

The research related to the structures of β-casein is contradicted and still unclear. Noelken et al. firstly predicted the form of β-casein in random coils in aqueous solution with only a small amount of regular structure [30], However, the results cannot be consistent due to the difference in temperature, ionic strength, protein concentration, etc., and these factors all contribute to affecting the ability of its self-aggregation [31]. β-casein exists as a monomer at low temperature and begins to aggregate as the temperature rises. While self-aggregation gradually increased and reached the maximum, once the protein concentration surpasses the threshold [32]. However, the effect of pH on the self-aggregation ability of β-casein is still inconclusive.

κ-casein is located on the surface of the casein micelle structure. Its most important role is to maintain the stability of the micelle structure and act as transition in the hydrophobic casein and water [33]. The unique disulfide bonds give κ-casein special properties different from the rest three types [34]. Though the self-aggregation of κ-casein consists of a core surrounded by multiple layers of variable polypeptide regions [35], which is similar to that of β-casein, it's not affected by temperature and ionic strength within a certain range and forms a fixed-size polymer.

In addition to nutritional functions, the potential biological activity of peptides formed as a result of casein proteolysis in the gastroenteric tract has been widely valued [36]. These peptides may affect the cardiovascular, nervous, immune, and digestive systems [37, 38]. However, it has been established that peptides are distributed unevenly in the composition of primary structure of caseins, which has attracted more attention to related research studies.

2.2 Whey proteins

Whey is a by-product during casein production, it can remain the form of liquid after coagulation by rennet, while the proteins left are the whey proteins [39]. Whey proteins are in full value and contain all kinds of amino acids that make up proteins [40]. And except for a little bit lower content in sulfur-containing amino acids, whey proteins have higher contents of the rest of essential amino acids compared with other proteins [41]. Therefore, whey contributes to nearly half of the nutrients in milk even though its total solids are only around 6.0 ~ 6.5% [16]. The high quality of whey proteins with complete and appropriate proportion of essential amino acids meets the requirement of human being, which also determines the functions of whey proteins. As mentioned above, whey proteins consist of a variety of biologically active ingredients including β-lactoglobulin, α-lactalbumin, serum albumin, immunoglobulin, and a variety of growth factors and biologically active peptides.

It's clear that the composition of amino acids determines the biological activity of each component in whey proteins. The various amino acids including threonine, cysteine, methionine are important to the intestine, muscle, and antioxidant systems [42]. While the physiological functions of whey proteins as well as the main amino acid related are reviewed below.

The β-lactoglobulin is the most prevalent one and comprises more than half of the whole proteins, while its prevalence affects the attributes of whey, β-lactoglobulin contains 162 amino acids with two variants differ in one amino acid, the disulfide and free sulfhydryl groups in its molecules forms let it become the major source of sulfur-contained amino acid [43]. β-lactoglobulin is manufactured in the ruminants while almost all of non-ruminants cannot synthesis it in the mammary gland [44]. The hydrophobic area on molecule of β-lactoglobulin is a quite effective in binding retinol, which may contribute to regulating the mammary gland by vitamin A [45]. However, the related biological functions of β-lactoglobulin are not commonly accepted now.

α-lactalbumin comprises around 13% of the whole whey proteins, with four disulfide linkages and no phosphate group molecule. The function of modifying the activity of the enzyme galactosyl transferase was proved in former studies, which promoting the transfer of UDP galactose to glucose [46]. α-lactalbumin is closely related to lysozyme evidenced by the similar synthesize of linkage but does not work on the same substrates, nor antigenically [47]. In addition, the α-lactalbumin is more heat-stable in the presence of calcium rather than in the absence of calcium [48], which is unusual compared with other proteins.

Different from the former two components, the serum albumin and immunoglobulins are not synthesized in the mammary gland. The serum albumins isolated from milk have the same molecule to those serum proteins since they are leaked into milk from the bloodstreams [49]. Therefore, serum albumins are identical to be the same molecule with serum proteins. Serum albumins contain no phosphorous with only one free sulfhydryl group, which gives those a specific binding sites for hydrophobic [50]. While the immunoglobulins in milk proteins comprise more than 2% of the total content and are classified by four types including IgG1, IgG2, IgA, and IgM. Immunoglobulins in colostrum can provide the passive immunity to the calf until the synthesis of antibodies activates in their body [51].

2.3 Milk fat globular membrane proteins

Milk fat globule membrane (MFGM) is the layer of film wrapped on the surface of milk fat, the function of which is to protect fat globule from polymerization or enzymatic degradation [52]. Milk fat globular membrane proteins (MFGMPs) are protein component in MFGM and contribute to the 25% ~ 70% of total contents [53]. MFGMPs have the most diverse biological functions and play an important role in the cell growth process and defense functions of newborns [54, 55].

The three-layer structure theory of milk fat globule membrane has been widely accepted [56]. While the main components of MFGMs including xanthine oxidoreductase, butyrophilin, and lactadherin play the most important role [57]. The innermost layer of the milk fat globule membrane is a single-layer membrane composed of polar lipids and proteins synthesized by the endoplasmic reticulum, which wraps the fat droplets in the core of the fat globule, followed by a high-density protein layer attached to the inner surface of the double-layer membrane; The last is the lipid bilayer that comes from the apical membrane of breast epithelial cells [58]. The cytoplasm forms a cytoplasmic crescent between the high-density protein layer and the outer double-membrane layer.

Xanthine oxidoreductase (XO/XDH) is the main component of milk fat globular membrane protein. It has been confirmed that it has a certain role in breast development, intestinal antibacterial and tissue damage [59]. XO/XDH is a member of the flavoprotein family of molybdenum dehydrogenase, which is a key enzyme for purine metabolism in the organism. However, the role of XO/XDH in the process of milk fat droplet wrapping and secretion may not express enzymes [60]. In addition, xanthine oxidoreductase has been confirmed to have a certain role in mammary gland development, intestinal antibacterial and tissue damage [61].

Butyrophilins (BTNs) are proteins related to fat droplets and a member of the immunoglobulin family. Many members of the BTN family have been confirmed to have immunomodulatory effects [62]. For example, the BTN3A1 and BTN3 families can inhibit T cell activation [63]. Milk-derived BTN has cross-reactivity with specific neuronal antibodies, which may be related to autoimmune regulation of diseases such as autism and multiple sclerosis [64].

Lactadherins are immunogenic lipophilic glycoproteins and are also known as milk fat globule epidermal growth factor 8 [65]. They are mainly distributed in secretory cells at the top of the milk tubules. In recent years, research on MFG-E8 has mainly focused on the phagocytosis of apoptotic cells, immune regulation, coagulation, and thrombosis [66]. Nakatani et al. found that MFG-E8 can recognize apoptotic cells in breast recession and activate phagocytes to phagocytose apoptotic epithelial cells. MFG-E8 can also resist rotavirus infection [67]. The protective effect of MFG-E8 on the intestinal tract has also become a research hotspot, which is mainly reflected in its anti-inflammatory, anti-apoptotic, and promoting intestinal mucosal repair effects [68].

3. Omics, a series of novel approaches to study milk protein

With the advent of the post-genomic era, the interactions between proteins, genes, and factors are followed by researchers, the diversity and dynamics of physiological functions cause the tradition approaches unable meet the requirement of current research studies. In recent years, a large number of research approaches including Omics have emerged in the research fields of biology [69, 70]. The emergence of these new applications can provide a complete view of the molecular regulation pathways of cells and organisms and make it possible to systematically investigate the lactation at the levels of genes, proteins, and even the metabolites, which is much helpful for the investigation of lactation especially the milk protein.

Omics are series of applications including genomics, transcriptomics, proteomics, and metabolomics in biological research. Genomics focus on the heterogeneity of coding genes, to investigate the sequence and expression of DNA, it provides the insight into the genetic structures by mapping as well as the performing the sequence analysis [71]. Transcriptomics profile the expression of mRNA in cells at specific time or state, while it can simultaneously work on more than thousands of changes in mRNA expression. Proteomics are used to determine the perturbations of expression patterns, abundance, and posttranslational modifications in proteins, and specialized in the differences caused by these factors [72]. While metabolomics monitor the changes in large groups of metabolites in biological samples, during which the further integration is conducted to reflect the physicochemical properties [73]. By applying these new approaches, the knowledge related to dairy science especially milk synthesis has been pushed forward tremendously in recent years, meanwhile the determination and analysis methods applied were developing as well, which were specifically overviewed in later section.

4. Genomics in milk protein research

Genomics involves genome mapping by genetic, physical, and transcript, gene sequence analysis and gene functional analysis, and was used in breeding selection. Since the DNA sequencing as well as high -density microarray analysis (gene chips) was commonly used in inferring genomics, the widespread of new next sequencing technology pushes the application of these two technologies in genomic studies related to lactation [74]. To improve the performance, dairy cows were already fully sequenced, and former studies of genomics in lactation research studies focused on the nutritional strategies and specific genes linked to the milk quality. However, the available data of genomics research on milk protein synthesis were still scarce in recent years. Since most of related studies concern more on the association of milk proteins with milk production traits instead of milk proteins variants with milk proteins composition [75]. The applications of genomics are mainly focused on the strong candidate of QTL, the variants of milk proteins, and the diseases that may affect the milk protein synthesis (i.e., Mastitis) [76]. Here we summarized those from mapping approaches to genome-wide association studies.

Lactation is usually affected by the factors such as environment, nutritional manages, breed, etc. The combination of the genetic data with the nutrition management including dry matter intake, body condition (in different periods) contributes to the efficient prediction of traits related to lactation as well as milk protein synthesis [77]. Former studies focused on the detection of relationships between lactation and the candidate genes related to milk proteins. And it has been shown that genes on chromosome 3 of bovine including the insulin-like growth factor 2 (IGF-2) and rap-1A are the strong candidate for the quantitative trait locus (QTL) and may affect the milk performance [78, 79], IGF-2 was found to have the functions of muscle mass and fat deposition in swine and was widely investigated in human medical. A parametric bootstrapping procedure found by Veerkamp et al. makes the estimate of heritability and genetic correlations between traits possible [80]. In addition, the development of genome-wide analysis makes it become a better solution to explain key genes and pathways. Berkowicz et al. indicated that a single genotyped single-nucleotide polymorphism (SNP) and traits related to animal growth also support the locus as harboring a potentially important quantitative trait nucleotides (QTNs) [79], suggesting that reprinter genes together with those documented biological roles represent important reservoirs for genetic improvement of dairy cows.

The genomics studies of milk proteins mainly focus on the amino acid changes caused by polymorphisms in the corresponding genes, while previous research studies focus on the variants on the components of caseins as well as whey proteins. To be specific, the A and B variants of β-LG and κ-CN on their concentration and total proteins in milk, which indicated that the A variant of β-LG is associated with a greater concentration of β-LG and a lesser concentration of casein, and that the B variant of κ-CN is associated with a greater concentration of κ-CN in milk [81].

But the research on the genotypes of all major components in milk proteins was not widely investigated in different kinds of animals. While in dairy cows, the A and B variants of β-LG and κ-CN, the E variant of κ-CN, and the A1, A2, and B variants of β-CN frequently occurred [82]. The genes of four different caseins have been found to be located at chromosome 6 of bovine and are closely linked meanwhile organized in a casein locus [83]. Heck et al. indicated that the selection for both the β-LG genotype B and the β-κ-CN haplotype A2B will result in cows that produce milk that is more suitable for production [84]. While related studies have also found the specific genes linked to milk proteins synthesis were also affected by the factors including seasons, lactation stages [81]. Therefore, the advance investigation of genes as well as variants will greatly enable the assessment of breeding to milk protein composition.

Mastitis is the most common disease in dairy cows, which is hallmarked by high somatic cell count (SCC). Mastitis negatively affects the dairy industry and enormously causes losses in milk performance and management costs [85]. Related research studies have shown that the stromal fibroblasts derived from cows that suffered mastitis were upregulated in the expression of inflammatory-related cytokines TNF-α and IL-8 [86], which contribute to the inhabitation the synthesis of milk proteins components including β-casein. Therefore, genomics research studies on mastitis have been conducted and SNPs on Bos Taurus autosome 4 (BTA4) and BTA18 were found to be significantly associated with mastitis [87]. Furthermore, 665 interactions in more than 140 genes were found in a co-expression network via the Genemania gene network analysis, which were recognized as candidate QTL for mastitis in the Holstein cows [88].

5. Transcriptomics in milk protein research

The transcriptome broadly refers to the sum of all RNAs transcribed by organism, tissues, or cells in a specific state or period. And its field can be classified by different RNA types or methodology [89]. Since gene expression is complex and involves a variety of regulatory mechanisms, such as histone modification, promoter region, variable shearing, all of which affect the intracellular genes expression and play a regulatory role [90]. The expression level of genes among different tissues or cells is not consistent; hence, the complex and variable functions can be performed during different condition of physiological and pathological [91].

Methods of transcriptomics are similar to genomics, including microarray and RNA sequencing, while the latter one overcomes the weakness of Sanger sequencing due to the higher accuracy as well as sensitivity. Nowadays, the next-generation platforms such as Illumia, SOLID, and 454 are commonly used in related research studies to obtain reads and the transcript assembled, which revolutionize the analysis of eukaryotic transcriptomic [92]. In research studies related to lactating dairy cows, the expression profiles of mRNA were more focused especially in mammary glands, to reveal the candidate genes related to lactation. While the long noncoding RNAs also contribute to the pathway regulations in mammary gland and are essential for breeding [93]. However, the quality as well as quantity of RNA determines the reliability of sequencing results, so the extraction of total RNA from mammary gland significantly affects the analysis of the transcriptome. Li et al. reviewed the RNA sources used in transcriptome of lactating bovine mammary gland [9], which were used to the investigate the mammary gland development, effects of nutritional management on synthesis of milk composition, etc.

To specifically investigate the synthesis of milk protein, former studies revealed that the insulin signaling and mammalian target of rapamycin (mTOR) pathway participate in the regulation of milk protein synthesis by RNA-seq [94]. MenziesK et al. demonstrated that insulin plays a pivotal role in regulating milk proteins in dairy cows, while the recent works also report the similar results in other mammal animals [95]. Insulin was found to both directly and indirectly regulate the milk protein synthesis that is the involvement of control in gene expression and the regulation in translation [96]. Specifically, insulin can strongly activate the STAT5 pathway by increasing phosphorylation of the transcription factor. While the expression of ELF5 is induced by insulin in mammary tissue of dairy cows, which can co-activate and amplify the STAT5 [97]. In addition, insulin can regulate amount of translation via the mTOR hence indirectly control milk protein synthesis. mTOR particularly MTORC1 among the two mTOR complexes was well defined to play an important role in regulate protein synthesis by affecting the translation in all tissues of mammals [94]. Protein synthesis in mammal is inhibited by the association of the unphosphorylated eukaryotic translation initiation factor 4E binding protein 1 (4EBP1) with the eukaryotic translation initiation factor 4E (eIF4E), the insulin can increase the specific phosphorylation in mTOR, which in turn phosphorylates 4EBP1 and results in releasing of eiF4E. eiF4E participates in the formation of the translation initiation complex and hence initiates the translation of mRNA into protein [98]. Furthermore, there are also additional mechanisms of activated mTOR complex to enhance the translation process through phosphorylation of the ribosomal protein S6 kinase (S6K1) and ukaryotic elongation factor-2 kinase (EEF2K) [99].

Except for the synthesis mechanism, transcriptomics conducted in investigation of nutritional management. The different expression of genes in milk synthesis was found in factors including short-term feeding restriction, low-quality total mixed ration, inducing the reduction in milk protein associated with the downregulation of protein synthesis [100, 101]. While the management such as the frequency of milking also causes the perturbation in lactation, former study reported that the increased milking frequency affects the expression of genes involved in reconstruction of extracellular matrix, κ-CN, α-lactalbumin, which may affect the milk protein components [102, 103]. Moreover, the related research studies also reported the perturbation of expression on milk protein synthesis induced by condition changes. For example, Gao et al. found that dairy cows suffered heat stress upregulated the inflammation-related genes, which interfered and downregulated the milk protein synthesis [104].

6. Proteomics in milk protein research

Proteomics is defined as the technology of protein expressions from transcription and translation. By analyzing the posttranslational modification meanwhile identifying the differential proteins [105], the interaction among proteins, proteomics reflects the body metabolic changes to internal and external environmental changes and reveals biological functions inside. The advancement of analytical techniques pushes the development of proteomics rapidly. Two-dimensional electrophoresis (2-DE) is a well-developed technology and was used to separate proteins in the past decade [106]. As early as 1977, O'Farrel used 2-DE technology to separate about 1100 proteins from *E. coli* [107], which proved that 2-DE technology has the characteristics of high resolution and high sensitivity and is effective for analyzing complex biological samples and separating proteins.

The 2-DE technology performs one-dimensional isoelectric focusing according to different isoelectric points of proteins and then separates proteins of different molecular weights by polyacrylamide gel electrophoresis [108]. The 2-DE formed by the combination of these two technologies effectively separates protein vesicles in both directions of charge and relative molecular mass. Therefore, the information of the differential protein can be obtained through subsequent mass spectrometry identification and combined with bioinformatics analysis to analyze its biological function. The application of 2-DE combined with MS was usually used to identify the biomarkers related to treatment [109]. However, quantitative proteomics is challenging the 2-DE and MS methods as it allows for the massive multiplexing of primary data with better quality than established methods [110]. The method of quantitative proteomics, the isobaric tags for absolute and relative quantification (iTRAQ) combined with MS, is commonly employed to analyze the multiplicity of different samples; however, the accuracy of iTRAQ might be compromised due to the influence of near isobaric ions contamination in a sample [111]. In fact, the methods of 2-DE combined with MS and iTRAQ have been widely selected in research related to lactating cows, particularly in the investigations of mastitis and effects of nutritional management.

As a novel research approach, proteomics was conducted to investigate the milk protein including the milk protein profile and the MFGM components. The investigation of lactation periods found the alternation of whey proteome as evidenced by the significantly decreased content of immunoglobulins and caseins and in particular the colostrum at 48 h postpartum [112]. While in mid-lactation, Reinhardt et al. found that the proteins associated with lipid transportation, synthesis, and secretion are upregulated, and 120 proteins were identified to associate with cell signaling and membrane/protein trafficking [113]. Except for the milk from different periods, the proteomic studies also focus on the milk components different among different species. For example, β-LG lacks in camel milk, while it is the main whey protein in the caprine and bovine [114]. This technology can also be used to identify the adulterated milk products meanwhile getting the information of the sources of hypoallergenic replacements. By the application of iTRAQ and MS, specific proteins from different species can be classified and be used for characterizing. For example, primary amine oxidase is unique in cows, while biglycan is source of goat [115, 116]. Furthermore, the investigation of MFGM profiles meanwhile provides the overview of MFGM proteome among species [117], which indicated that these differences of protein components may be related to differences in heredity.

The application of proteomics in pathology such as mastitis has been conducted to find biomarkers to overcome the challenges of quantification in mastitis diagnosing. While proteins including hemoglobin β, cytochrome C oxidase, annexin V, and α-1-acid glycoprotein as well as collagen type I α 1 and inter-α (Globulin) inhibitor H4 show more abundance in dairy cows that suffered mastitis [118, 119], which may participate in the repairment of tissue damage.

7. Metabolomics in milk protein research

Metabolomics focus on the quantitatively analyzing of all metabolites in body to find the relative relationship between metabolites and physiological changes. Most of its research objects are small molecular substances with a relative molecular mass of less than 1000. In recent years, studies have reported the effects of different environmental factors on small molecular metabolites in animals and plants. While metabolic map was drawn through in-depth research on the metabolites of the body to seek the biomarkers [120, 121]. Moreover, analysis of NMR and MS indicated that biomarkers such as phosphorylated saccharides, acetone, and β-hydroxybutyrate (high levels) are closely correlated with the metabolic status in dairy cows during early lactation [9], which may contribute to the breeding selection to alleviate the metabolic stress in dairy cows during early lactation.

Similar to the application of proteomics in searching unique biomarkers, the investigation using NMR metabolomics approach found that acetate and novel metabolites including hippurate, isoleucine, butyrate, fumarate, and β-hydroxybutyrate are associated with milk composition [122]. While improving abundance of volatile fatty acid (VFA) as well as β-hydroxybutyrate and low abundance of hippurate and fumarate in milk are coupled with high levels of somatic cells [123], which contribute to the potential biomarkers for milk quality when dairy cows in high SCC conditions. Furthermore, metabolomics also provides new approach on highlighting interspecies differences from analyzing the metabolites. By using NMR and LC–MS or GC–MS method, the unique metabolites in milk among horse, Jersey cow, camel, yak, goat, caprine, buffalo, and dairy cow were found respectively, and results validated that metabolomics is a feasible approach for milk composition analysis [124, 125].

Although metabolomics has a higher accuracy and the high-throughput abilities, its application is still in the junior stages in research studies related to lactation especially the milk protein. With the advantage of combination with multivariate data analysis tool, metabolomics will obviously push the process of lactation research in the future. While the unique metabolites identified by these technologies will also provide much better perspective for the investigation of milk protein.

8. Conclusions

The synthesis of milk protein is complex in mammary gland and is regulated by multiple factors, the protein components in milk are specific with the mammary tissue since they are minor or no expressions in the rest organs. Several hormones including insulin participate in the regulation of milk protein synthesis with a pivotal role of perturbation pathway such as mTOR. However, the control of milk synthesis is still need to be completely understood, to open new insight of research not previously considered. The advent of Omics technologies provides the possibilities to further investigate related complex mechanisms. These novel research approaches combined with bioinformatics constitute to be useful and powerful to generate large datasets for lactation sciences, which contribute to reveal the mechanism of milk protein synthesis and the novel biomarkers in milk affected by some factors. However, considering the limitation of cost, reproducibility, and throughput, it should be well arranged and prepared when choosing these new research approaches.

Funding

The present study was financially supported by the National Natural Science Foundation of China (31872383), the Agriculture Science and Technology Innovation Program (ASTIP-IAS07-1) and the Key Research and Development Program of the Ningxia Hui Autonomous Region (2021BEF02018).

Author details

Zitai Guo, Lu Ma and Dengpan Bu*
State Key Laboratory of Animal Nutrition, Institute of Animal Sciences,
Chinese Academy of Agricultural Sciences, Beijing, China

*Address all correspondence to: budengpan@126.com

References

[1] Ha E, Zemel MB. Functional properties of whey, whey components, and essential amino acids: Mechanisms underlying health benefits for active people (review). Journal of Nutritional Biochemistry. 2003;**14**(5):251-258

[2] Guo Z, Gao S, Ouyang J, Ma L, Bu D. Impacts of heat stress-induced oxidative stress on the Milk protein biosynthesis of dairy cows. Animals. 2021;**11**(3):726

[3] Rhoads ML, Rhoads RP, VanBaale MJ, Collier RJ, Sanders SR, Weber WJ, et al. Effects of heat stress and plane of nutrition on lactating Holstein cows: I. production, metabolism, and aspects of circulating somatotropin. Journal of Dairy Science. 2009;**92**(5):1986-1997

[4] Bionaz M, Loor JJ. Gene networks driving bovine milk fat synthesis during the lactation cycle. BMC Genomics. 2008;**9**(1):366

[5] Ayadi M, Caja G, Such X, Knight CH. Effect of omitting one milking weekly on lactational performances and morphological udder changes in dairy cows. Journal of Dairy Science. 2003;**86**(7):2352-2358

[6] Gao ST, Guo J, Quan SY, Nan XM, Fernandez MVS, Baumgard LH, et al. The effects of heat stress on protein metabolism in lactating Holstein cows. Journal of Dairy Science. 2017;**100**(6):5040-5049

[7] Nooren IM, Thornton JM. Diversity of protein-protein interactions. The EMBO Journal. 2003;**22**(14):3486-3492

[8] Vailati-Riboni M, Palombo V, Loor JJ. What are omics sciences? In: Periparturient Diseases of Dairy Cows. Switzerland: Springer; 2017. pp. 1-7

[9] Li S, Wang Q, Lin X, Jin X, Liu L, Wang C, et al. The use of "omics" in lactation research in dairy cows. International Journal of Molecular Sciences. 2017;**18**(5):983

[10] Szwajkowska M, Wolanciuk A, Barłowska J, Krol J, Litwińczuk Z. Bovine milk proteins as the source of bioactive peptides influencing the consumers' immune system–a review. Animal Science Papers and Reports. 2011;**29**(4):269-280

[11] NG-Kwai-Hang KF, Hayes JF, Moxley JE, Monardes HG. Variation in milk protein concentrations associated with genetic polymorphism and environmental factors. Journal of Dairy Science. 1987;**70**(3):563-570

[12] Livney YD. Milk proteins as vehicles for bioactives. Current Opinion in Colloid & Interface Science. 2010;**15**(1-2):73-83

[13] De Wit JN. Marschall Rhone-Poulenc award lecture. Nutritional and functional characteristics of whey proteins in food products. Journal of Dairy Science. 1998;**81**(3):597-608

[14] Azarpazhooh A, Limeback H. Clinical efficacy of casein derivatives: A systematic review of the literature. Journal of the American Dental Association. 2008;**139**(7):915-924. quiz 994-5

[15] Hipp N, Groves ML, Custer J, McMeekin T. Separation of α-, β-and γ-casein. Journal of Dairy Science. 1952;**35**(3):272-281

[16] Kilara A, Vaghela M. Whey proteins. In: Proteins in Food Processing. Sawston, UK: Elsevier; 2018. pp. 93-126

[17] Demmelmair H, Prell C, Timby N, Lonnerdal B. Benefits of Lactoferrin, Osteopontin and Milk fat globule membranes for infants. Nutrients. 2017;**9**(8):817

[18] Cross KJ, Huq NL, Palamara JE, Perich JW, Reynolds EC. Physicochemical characterization of casein phosphopeptide-amorphous calcium phosphate nanocomplexes. Journal of Biological Chemistry. 2005;**280**(15):15362-15369

[19] Molino ACT, Aline C. Effects of Enzymatic Dephosphorylation on Properties of Bovine Casein. Canada, Ottawa: McGill University; 2006 [dissertation]

[20] McMahon DJ, Brown RJ. Composition, structure, and integrity of casein micelles: A review. Journal of Dairy Science. 1984;**67**(3):499-512

[21] McMahon DJ, Oommen BS. Supramolecular structure of the casein micelle. Journal of Dairy Science. 2008;**91**(5):1709-1721

[22] Plank J, Bian H. Method to assess the quality of casein used as superplasticizer in self-levelling compounds. Cement and Concrete Research. 2010;**40**(5):710-715

[23] Horne DS. A balanced view of casein interactions. Current Opinion in Colloid & Interface Science. 2017;**28**:74-86

[24] Bigelow CC. On the average hydrophobicity of proteins and the relation between it and protein structure. Journal of Theoretical Biology. 1967;**16**(2):187-211

[25] Heinecke JW, Rosen H, Suzuki LA, Chait A. The role of sulfur-containing amino acids in superoxide production and modification of low density lipoprotein by arterial smooth muscle cells. Journal of Biological Chemistry. 1987;**262**(21):10098-10103

[26] Orsini S, Parlanti F, Bonaduce I. Analytical pyrolysis of proteins in samples from artistic and archaeological objects. Journal of Analytical and Applied Pyrolysis. 2017;**124**:643-657

[27] Liang L, Luo Y. Casein and pectin: Structures, interactions, and applications. Trends in Food Science and Technology. 2020;**97**:391-403

[28] Horne DS. Casein interactions: Casting light on the black boxes, the structure in dairy products. International Dairy Journal. 1998;**8**(3):171-177

[29] Swaisgood HE. Chemistry of milk protein. Developments in Dairy Chemistry. 1982;**1**:1-59

[30] Noelken M, Reibstein M. Conformation of beta-casein B. Archives of Biochemistry and Biophysics. 1968;**123**(2):397-402

[31] Wang W, Nema S, Teagarden D. Protein aggregation--pathways and influencing factors. International Journal of Pharmaceutics. 2010;**390**(2):89-99

[32] Tai M, Kegeles G. A micelle model for the sedimentation behavior of bovine beta-casein. Biophysical Chemistry. 1984;**20**(1-2):81-87

[33] Creamer LK, Plowman JE, Liddell MJ, Smith MH, Hill JP. Micelle stability: Kappa-casein structure and function. Journal of Dairy Science. 1998;**81**(11):3004-3012

[34] Groves, M. L.; Wickham, E. D.; Farrell, H. M., Jr., Environmental effects on disulfide bonding patterns of bovine kappa-casein. Journal of Protein Chemistry 1998, 17 (2), 73-84.

[35] Phadungath C. Casein micelle structure: a concise review. Songklanakarin Journal of Science and Technology. 2005;**27**(1):201-212

[36] Yukalo V, Storozh L. Isolation of κ-CN-1P and β-CN-5P fractions from native casein micelles. The Ukrainian. Biochemical Journal. 2018;**90**(4):74-79

[37] Altmann K, Wutkowski A, Klempt M, Clawin-Rädecker I, Meisel H, Lorenzen PC. Generation and identification of anti-inflammatory peptides from bovine β-casein using enzyme preparations from cod and hog. Journal of the Science of Food and Agriculture. 2016;**96**(3):868-877

[38] Myronovskij S, Negrych N, Nehrych T, Starykovych M, Tkachenko V, Yukalo V, et al. Isolation and characterization of peptides from blood serum of patients with multiple sclerosis. Studia. Biologicals. 2015;**9**(2):51-58

[39] Smithers GW. Whey and whey proteins—From 'gutter-to-gold'. International Dairy Journal. 2008;**18**(7):695-704

[40] Luhovyy BL, Akhavan T, Anderson GH. Whey proteins in the regulation of food intake and satiety. Journal of the American College of Nutrition. 2007;**26**(6):704S-712S

[41] Csapó-Kiss Z, Stefler J, Martin T, Makray S, Csapó J. Composition of mares' colostrum and milk. Protein content, amino acid composition and contents of macro and micro-elements. International Dairy Journal. 1995;**5**(4):403-415

[42] Liu Y, Wang X, Hu CA. Therapeutic potential of amino acids in inflammatory bowel disease. Nutrients. 2017;**9**(9):920

[43] Fox PF, Mulvihill DM. Milk proteins: Molecular, colloidal and functional properties. Journal of Dairy Research. 1982;**49**(4):679-693

[44] Pérez MD, Calvo M. Interaction of β-lactoglobulin with retinol and fatty acids and its role as a possible biological function for this protein: A review. Journal of Dairy Science. 1995;**78**(5):978-988

[45] Sawyer L. β-Lactoglobulin. In: Advanced Dairy Chemistry. New York: Springer; 2013. pp. 211-259

[46] Morrison J, Ebner K. Studies on galactosyltransferase: Kinetic effects of α-lactalbumin with N-acetylglucosamine and glucose as galactosyl group acceptors. Journal of Biological Chemistry. 1971;**246**(12):3992-3998

[47] Kilara A. Whey and whey products. Dairy Processing Quality Assurance. 2008;**1**:337-355

[48] Nielsen LR, Nielsen SB, Zhao Z, Olsen K, Nielsen JH, Lund MN. Control of alpha-Lactalbumin aggregation by modulation of temperature and concentration of calcium and cysteine. Journal of Agricultural and Food Chemistry. 2018;**66**(27):7110-7120

[49] Shamay A, Homans R, Fuerman Y, Levin I, Barash H, Silanikove N, et al. Expression of albumin in nonhepatic tissues and its synthesis by the bovine mammary gland. Journal of Dairy Science. 2005;**88**(2):569-576

[50] Spector AA, John K, Fletcher JE. Binding of long-chain fatty acids to bovine serum albumin. Journal of Lipid Research. 1969;**10**(1):56-67

[51] Butler JE. Bovine immunoglobulins: A review. Journal of Dairy Science. 1969;**52**(12):1895-1909

[52] Manion B. Compositional and Functional Properties of Fractions Rich in Milk Fat Globule Membrane Material. Ontario, Canada: University of Guelph; 2007

[53] Danthine S, Blecker C, Paquot M, Innocente N, Deroanne C. Progress in milk fat globule membrane research: A review. Le Lait. 2000;**80**(2):209-222

[54] Cavaletto M, Giuffrida MG, Conti A. Milk fat globule membrane components--a proteomic approach. Advances in Experimental Medicine and Biology. 2008;**606**:129-141

[55] Zavaleta N, Kvistgaard AS, Graverholt G, Respicio G, Guija H, Valencia N, et al. Efficacy of an MFGM-enriched complementary food in diarrhea, anemia, and micronutrient status in infants. Journal of Pediatric Gastroenterology and Nutrition. 2011;**53**(5):561-568

[56] Dewettinck K, Rombaut R, Thienpont N, Le TT, Messens K, Van Camp J. Nutritional and technological aspects of milk fat globule membrane material. International Dairy Journal. 2008;**18**(5):436-457

[57] Keenan TW. Historical perspective: Milk lipid globules and their surrounding membrane: A brief history and perspectives for future research. Journal of Mammary Gland Biology and Neoplasia. 2001;**6**(3):365-371

[58] Evers JM. The milkfat globule membrane—Compositional and structural changes post secretion by the mammary secretory cell. International Dairy Journal. 2004;**14**(8):661-674

[59] Mather IH. A review and proposed nomenclature for major proteins of the milk-fat globule membrane. Journal of Dairy Science. 2000;**83**(2):203-247

[60] Nishino T, Okamoto K, Eger BT, Pai EF. 5 the xanthine oxidoreductase enzyme family: Xanthine dehydrogenase, xanthine oxidase, and aldehyde oxidase. In: Volume 2 Complex Flavoproteins, Dehydrogenases and Physical Methods, De Gruyter. Berlin, Boston: Walter de Gruyter GmbH; 2013. pp. 103-124. Available from: https://books.google.com.hk/books?hl=en&lr=&id=5a_pBQAAQBAJ&oi=fnd&pg=PR5&dq=Handbook+of+Flavoproteins:+Complex+Flavoproteins,+Dehydrogenases+and+Physical+Methods&ots=vYfyogkivI&sig=_xNVrJkkj-1fEq87STwOJsUBUG0&redir_esc=y#v=onepage&q=Handbook%20of%20Flavoproteins%3A%20Complex%20Flavoproteins%2C%20Dehydrogenases%20and%20Physical%20Methods&f=false

[61] Martin HM, Hancock JT, Salisbury V, Harrison R. Role of xanthine oxidoreductase as an antimicrobial agent. Infection and Immunity. 2004;**72**(9):4933-4939

[62] Abeler-Dorner L, Swamy M, Williams G, Hayday AC, Bas A. Butyrophilins: An emerging family of immune regulators. Trends in Immunology. 2012;**33**(1):34-41

[63] Karunakaran MM, Herrmann T. The Vγ9Vδ2 T cell antigen receptor and butyrophilin 3 A1 (BTN3A1): Models of interaction, the possibility of co-evolution, and the case of dendritic epidermal T cells. Frontiers in Immunology. 2014;**5**:648

[64] Borba VV, Lerner A, Matthias T, Shoenfeld Y. Bovine milk proteins as a trigger for autoimmune diseases: Myth or reality? International Journal. 2020;**8**(1):10-21

[65] Sun Y, Wang C, Sun X, Guo M. Comparative proteomics of whey and Milk fat globule membrane proteins of Guanzhong goat and Holstein cow mature Milk. Journal of Dairy Science. 2019;**84**(2):244-253

[66] Aziz M, Jacob A, Matsuda A, Wang P. Review: Milk fat globule-EGF factor 8 expression, function and plausible signal transduction in resolving inflammation. Apoptosis. 2011;**16**(11):1077-1086

[67] Nakatani H, Aoki N, Nakagawa Y, Jin-No S, Aoyama K, Oshima K, et al. Weaning-induced expression of a milk-fat globule protein, MFG-E8, in mouse mammary glands, as demonstrated by the analyses of its mRNA, protein and phosphatidylserine-binding activity. Biochemical Journal. 2006;**395**(1):21-30

[68] Peterson JA, Scallan CD, Ceriani RL, Hamosh M. Structural and functional aspects of three major glycoproteins of the human milk fat globule membrane. Advances in Experimental Medicine and Biology. 2001;**501**:179-187

[69] Cranford SW, De Boer J, Van Blitterswijk C, Buehler MJ. Materiomics: An-omics approach to biomaterials research. Advanced Materials. 2013;**25**(6):802-824

[70] Rochfort S. Metabolomics reviewed: A new "omics" platform technology for systems biology and implications for natural products research. Journal of Natural Products. 2005;**68**(12):1813-1820

[71] Bustamante CD, Burchard EG, De la Vega FM. Genomics for the world. Nature. 2011;**475**(7355):163-165

[72] Hegde PS, White IR, Debouck C. Interplay of transcriptomics and proteomics. Current Opinion in Biotechnology. 2003;**14**(6):647-651

[73] Dettmer K, Aronov PA, Hammock BD. Mass spectrometry-based metabolomics. Mass Spectrometry Reviews. 2007;**26**(1):51-78

[74] Varshney RK, Graner A, Sorrells ME. Genomics-assisted breeding for crop improvement. Trends in Plant Science. 2005;**10**(12):621-630

[75] Amigo L, Recio I, Ramos M. Genetic polymorphism of ovine milk proteins: Its influence on technological properties of milk—A review. International Dairy Journal. 2000;**10**(3):135-149

[76] Barillet F. Genetic improvement for dairy production in sheep and goats. Small Ruminant Research. 2007;**70**(1):60-75

[77] Bilal G, Cue R, Hayes J. Genetic and phenotypic associations of type traits and body condition score with dry matter intake, milk yield, and number of breedings in first lactation Canadian Holstein cows. Canadian Journal of Animal Science. 2016;**96**(3):434-447

[78] Cohen-Zinder M, Donthu R, Larkin DM, Kumar CG, Rodriguez-Zas SL, Andropolis KE, et al. Multisite haplotype on cattle chromosome 3 is associated with quantitative trait locus effects on lactation traits. Physiological Genomics. 2011;**43**(21):1185-1197

[79] Berkowicz EW, Magee DA, Sikora KM, Berry DP, Howard DJ, Mullen MP, et al. Single nucleotide polymorphisms at the imprinted bovine insulin-like growth factor 2 (IGF2) locus are associated with dairy performance in Irish Holstein-Friesian cattle. Journal of Dairy Research. 2011;**78**(1):1-8

[80] Veerkamp R, Mulder H, Thompson R, Calus M. Genomic and pedigree-based genetic parameters for scarcely recorded traits when some

animals are genotyped. Journal of Dairy Science. 2011;**94**(8):4189-4197

[81] Bobe G, Beitz DC, Freeman AE, Lindberg GL. Effect of milk protein genotypes on milk protein composition and its genetic parameter estimates. Journal of Dairy Science. 1999;**82**(12):2797-2804

[82] Boettcher PJ, Caroli A, Stella A, Chessa S, Budelli E, Canavesi F, et al. Effects of casein haplotypes on milk production traits in Italian Holstein and Brown Swiss cattle. Journal of Dairy Science. 2004;**87**(12):4311-4317

[83] Li L. Physical Mapping of Two Multi-Gene Families and Comparative Anchor Loci in Cattle and Genetic Linkage Analysis on Cattle Chromosome 6. Texas A&M University; 1993 [phD thesis]

[84] Heck J, Olieman C, Schennink A, Van Valenberg H, Visker M, Meuldijk R, et al. Estimation of variation in concentration, phosphorylation and genetic polymorphism of milk proteins using capillary zone electrophoresis. International Dairy Journal. 2008;**18**(5):548-555

[85] Klopfenstein JJ. Dairy herd health for optimal reproduction. Bovine Reproduction. 2021:517-525

[86] Moyes KM, Drackley JK, Morin DE, Bionaz M, Rodriguez-Zas SL, Everts RE, et al. Gene network and pathway analysis of bovine mammary tissue challenged with streptococcus uberis reveals induction of cell proliferation and inhibition of PPARγ signaling as potential mechanism for the negative relationships between immune response and lipid metabolism. BMC Genomics. 2009;**10**(1):1-29

[87] Wijga S, Bastiaansen JW, Wall E, Strandberg E, de Haas Y, Giblin L, et al. Genomic associations with somatic cell score in first-lactation Holstein cows. Journal of Dairy Science. 2012;**95**(2):899-908

[88] Tiezzi F, Parker-Gaddis KL, Cole JB, Clay JS, Maltecca C. A genome-wide association study for clinical mastitis in first parity US Holstein cows using single-step approach and genomic matrix re-weighting procedure. PLoS One. 2015;**10**(2):e0114919

[89] Lowe R, Shirley N, Bleackley M, Dolan S, Shafee T. Transcriptomics technologies. PLoS Computational Biology. 2017;**13**(5):e1005457

[90] Karlic R, Chung HR, Lasserre J, Vlahovicek K, Vingron M. Histone modification levels are predictive for gene expression. Proceedings of the National Academy of Sciences of the United States of America. 2010;**107**(7):2926-2931

[91] Perou CM, Jeffrey SS, van de Rijn M, Rees CA, Eisen MB, Ross DT, et al. Distinctive gene expression patterns in human mammary epithelial cells and breast cancers. Proceedings of the National Academy of Sciences of the United States of America. 1999;**96**(16):9212-9217

[92] Mutz K-O, Heilkenbrinker A, Lönne M, Walter J-G, Stahl F. Transcriptome analysis using next-generation sequencing. Current Opinion in Biotechnology. 2013;**24**(1):22-30

[93] Yang B, Jiao B, Ge W, Zhang X, Wang S, Zhao H, et al. Transcriptome sequencing to detect the potential role of long non-coding RNAs in bovine mammary gland during the dry and lactation period. BMC Genomics. 2018;**19**(1):605

[94] Wang X, Proud CG. The mTOR pathway in the control of protein

synthesis. Physiology (Bethesda). 2006;**21**(5):362-369

[95] Menzies KK, Lefèvre C, Macmillan KL, Nicholas KR. Insulin regulates milk protein synthesis at multiple levels in the bovine mammary gland. Functional & Integrative Genomics. 2009;**9**(2):197-217

[96] Cao Y, Liu S, Liu K, Abbasi IHR, Cai C, Yao J. Molecular mechanisms relating to amino acid regulation of protein synthesis. Nutrition Research Reviews. 2019;**32**(2):183-191

[97] Sandhu GK, Milevskiy MJG, Wilson W, Shewan AM, Brown MA. Non-coding RNAs in mammary gland development and disease. Advances in Experimental Medicine and Biology. 2016;**886**:121-153

[98] Kimball SR, Jefferson LS. New functions for amino acids: Effects on gene transcription and translation. The American Journal of Clinical Nutrition. 2006;**83**(2):500S-507S

[99] Schieke, S. M.; Phillips, D.; McCoy, J. P., Jr.; Aponte, A. M.; Shen, R. F.; Balaban, R. S.; Finkel, T., The mammalian target of rapamycin (mTOR) pathway regulates mitochondrial oxygen consumption and oxidative capacity. Journal of Biological Chemistry 2006, 281 (37), 27643-27652.

[100] Ollier S, Robert-Granie C, Bernard L, Chilliard Y, Leroux C. Mammary transcriptome analysis of food-deprived lactating goats highlights genes involved in milk secretion and programmed cell death. Journal of Nutrition. 2007;**137**(3):560-567

[101] Sigl T, Meyer HH, Wiedemann S. Gene expression analysis of protein synthesis pathways in bovine mammary epithelial cells purified from milk during lactation and short-term restricted feeding. Journal of Animal Physiology and Animal Nutrition. 2014;**98**(1):84-95

[102] Erdman RA, Varner M. Fixed yield responses to increased milking frequency. Journal of Dairy Science. 1995;**78**(5):1199-1203

[103] Boutinaud M, Ben Chedly MH, Delamaire E, Guinard-Flament J. Milking and feed restriction regulate transcripts of mammary epithelial cells purified from milk. Journal of Dairy Science. 2008;**91**(3):988-998

[104] Gao S, Ma L, Zhou Z, Zhou Z, Baumgard LH, Jiang D, et al. Heat stress negatively affects the transcriptome related to overall metabolism and milk protein synthesis in mammary tissue of midlactating dairy cows. Physiological Genomics. 2019;**51**(8):400-409

[105] Pappireddi N, Martin L, Wuhr M. A review on quantitative multiplexed proteomics. Chembiochem. 2019;**20**(10):1210-1224

[106] Ye X, Tang J, Mao Y, Lu X, Yang Y, Chen W, et al. Integrated proteomics sample preparation and fractionation: Method development and applications. TrAC Trends in Analytical Chemistry. 2019;**120**:115667

[107] O'Farrell PZ, Goodman HM, O'Farrell PH. High resolution two-dimensional electrophoresis of basic as well as acidic proteins. Cell. 1977;**12**(4):1133-1142

[108] Issaq H, Veenstra T. Two-dimensional polyacrylamide gel electrophoresis (2D-PAGE): Advances and perspectives. BioTechniques. 2008;**44**(5):697-698, 700

[109] Chen J, Zhou C, Liu Z, Fu Y, Zheng P, Yang D, et al. Divergent urinary metabolic phenotypes between major

depressive disorder and bipolar disorder identified by a combined GC–MS and NMR spectroscopic metabonomic approach. Journal of Proteome Research. 2015;**14**(8):3382-3389

[110] Bantscheff M, Lemeer S, Savitski MM, Kuster B. Quantitative mass spectrometry in proteomics: Critical review update from 2007 to the present. Analytical and Bioanalytical Chemistry. 2012;**404**(4):939-965

[111] Bantscheff M, Schirle M, Sweetman G, Rick J, Kuster B. Quantitative mass spectrometry in proteomics: A critical review. Analytical and Bioanalytical Chemistry. 2007;**389**(4):1017-1031

[112] Senda A, Fukuda K, Ishii T, Urashima T. Changes in the bovine whey proteome during the early lactation period. Animal Science Journal. 2011;**82**(5):698-706

[113] Reinhardt TA, Lippolis JD. Developmental changes in the milk fat globule membrane proteome during the transition from colostrum to milk. Journal of Dairy Science. 2008;**91**(6):2307-2318

[114] Hinz K, O'Connor PM, Huppertz T, Ross RP, Kelly AL. Comparison of the principal proteins in bovine, caprine, buffalo, equine and camel milk. Journal of Dairy Research. 2012;**79**(2):185-191

[115] Yang Y, Bu D, Zhao X, Sun P, Wang J, Zhou L. Proteomic analysis of cow, yak, buffalo, goat and camel milk whey proteins: Quantitative differential expression patterns. Journal of Proteome Research. 2013;**12**(4):1660-1667

[116] Hettinga K, van Valenberg H, de Vries S, Boeren S, van Hooijdonk T, van Arendonk J, et al. The host defense proteome of human and bovine milk. PLoS One. 2011;**6**(4):e19433

[117] Yang Y, Zheng N, Zhao X, Zhang Y, Han R, Ma L, et al. Proteomic characterization and comparison of mammalian milk fat globule proteomes by iTRAQ analysis. Journal of Proteomics. 2015;**116**:34-43

[118] Eckersall PD, McLaughlin M, Eckersall P, Whitfield P. 10 proteomics in animal health and disease. Methods in Animal Proteomics. 2011;**243**:5-11

[119] Mansor R. Proteomic and Metabolomic Studies on Milk during Bovine Mastitis. Scotland, UK: University of Glasgow; 2012

[120] Garcia-Sevillano MA, Contreras-Acuna M, Garcia-Barrera T, Navarro F, Gomez-Ariza JL. Metabolomic study in plasma, liver and kidney of mice exposed to inorganic arsenic based on mass spectrometry. Analytical and Bioanalytical Chemistry. 2014;**406**(5):1455-1469

[121] Wei L, Wang Q, Wu H, Ji C, Zhao J. Proteomic and metabolomic responses of Pacific oyster Crassostrea gigas to elevated pCO2 exposure. Journal of Proteomics. 2015;**112**:83-94

[122] Sundekilde U, Poulsen NA, Larsen LB, Bertram H. Nuclear magnetic resonance metabonomics reveals strong association between milk metabolites and somatic cell count in bovine milk. Journal of Dairy Science. 2013;**96**(1):290-299

[123] Boudonck KJ, Mitchell MW, Wulff J, Ryals JA. Characterization of the biochemical variability of bovine milk using metabolomics. Metabolomics. 2009;**5**(4):375-386

[124] Scano P, Murgia A, Pirisi FM, Caboni P. A gas chromatography-mass spectrometry-based metabolomic approach for the characterization of goat milk compared with cow milk. Journal of Dairy Science. 2014;**97**(10):6057-6066

[125] Yang Y, Zheng N, Zhao X, Zhang Y, Han R, Yang J, et al. Metabolomic biomarkers identify differences in milk produced by Holstein cows and other minor dairy animals. Journal of Proteomics. 2016;**136**:174-182

Section 4

Milk Protein and Food Allergens

Chapter 5

Characteristics of Cow Milk Proteins and the Effect of Processing on Their Allergenicity

Roua Lajnaf, Sawsan Feki, Hamadi Attia, Mohamed Ali Ayadi and Hatem Masmoudi

Abstract

Milk proteins are well known for their nutritional and functional properties. However, they are also members of the Big-8 food allergens including egg, fish, shellfish, soy, peanuts, wheat and tree nuts, in terms of prevalence. The most common milk allergens are casein fractions and β-lactoglobulin naturally not present in human breast milk. Thus, the examination of cow's milk proteins as potential allergens that may cause food allergies and the identification of methods of reducing their immunogenicity are of great interest. The main objective of this chapter is to review the physico-chemical characteristics cow milk proteins as well as their studied allergenicity and immunogenicity as a function of some denatured dairy processes such as heating, high pressure, enzymatic hydrolysis and lactic acid fermentation.

Keywords: cow's milk proteins allergy, protein allergenicity, immunoreactivity, milk processing, β-lactoglobulin, caseins

1. Introduction

Food allergy is a major public health which has been estimated to affect around 1–2% of the adult population and 5–8% of pediatric population at the age below 3 years [1–3]. It is thought to result from disorders of the immune response to food allergens proteins and develop due to the defect in oral tolerance. Food allergens are contained in eight common foods including animal-based foods (cow's milk, eggs and fish) and plant-based foods (crustacean/shellfish, peanuts, soy, nuts and wheat). These allergens account for over 90% of the occurrence of all serious allergic reactions to foods worldwide [4]. Epidemiological studies have reported that animal food allergens, especially cow milk proteins allergy, was the most prevalent allergy for infants or young children, meanwhile, plant based food allergens was more encountered for adults [5].

Thus, Cow's Milk Allergy (CMA) represents 10–40% of the total food allergies. As an animal proteins allergy, it concerns mostly young children and less frequently adults. CMA is reported to affect approximately 3–8% of the total pediatric population worldwide with symptoms at different levels of severity, which can endanger the patient's life [6]. Indeed, CMA is considered as the most common food allergy responsible for anaphylaxis reactions in young children as it is the first food eaten since birth. CMA is ranked third among all food allergies responsible for serious anaphylactic reactions to adults representing 15% from all allergic cases [7].

CMA disappears spontaneously at the age of 5 years in the majority of patients representing approximately 80% [8]. However, it seems that a minority of patients remains allergic in adulthood [9]. In most cases, the food allergy manifests itself as an immediate hypersensitivity reaction induced after recognition of food antigens by specific immunoglobulins type E (IgE). Other forms of allergy can also involve mechanisms not mediated by IgE. Their frequency is increasing but the immune responses involved are still poorly defined.

Like all food allergies, CMA involves both immunological reactions: immunoglobulin E (IgE) which is encountered in most allergic cases and non-IgE mediated reactions [10, 11]. The immunological reactions of IgE-mediated reactions occur immediately after proteins ingestion because of the interaction between allergens and immune mechanisms. This allergenic reaction is characterized by the production of IgE antibodies in allergic patients resulting in the degranulation phenomena of mast cells, the release of inflammatory mediators including histamine, 5-hydroxytryptamine (5HT) and prostaglandin E2 (PGE2) (**Figure 1**). These mediators induced the resulting allergy symptoms (hives, diarrhea, vomiting, and breathing difficulty). On the other hand, non-IgE mediated immunological reactions take up between 1hour and several days after ingestion of milk to develop involving the immune system as the IgE-mediated reactions [12–14].

The complete exclusion of cow's milk protein from the diet is still the only safe treatment that can be offered to patients today. In this case, infant formulations

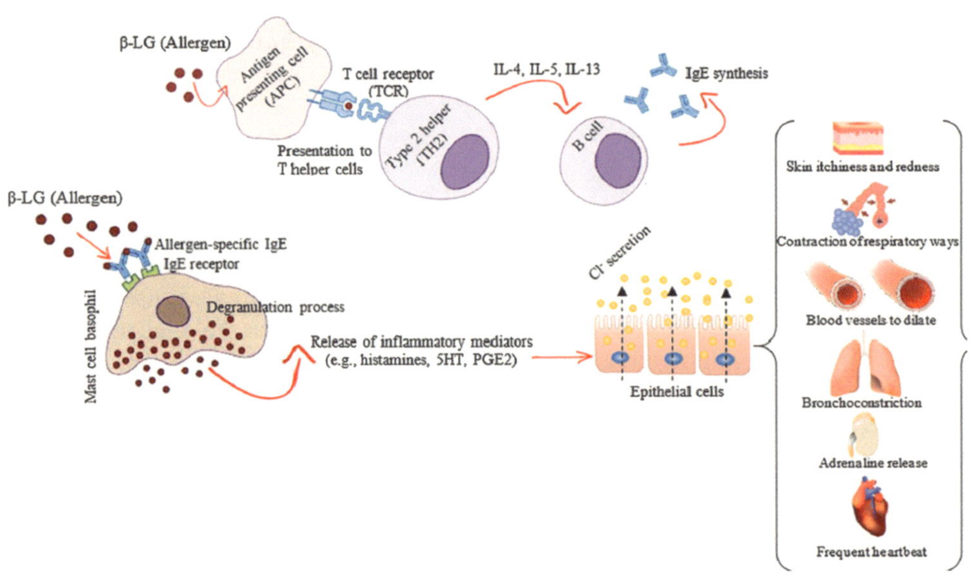

Figure 1.
A schematic representation of allergenicity mechanism of bovine milk in the body (abbreviations: β-Lg: β-lactoglobulin; IL-4, IL-5 and IL-13, inflammatory cytokines; IgE, immunoglobulin E; 5HT, 5-hydroxytryptamine; PGE2, prostaglandin E2) [12].

containing cow's milk proteins are replaced by milks which are designed from more or less extensive hydrolysates of bovine proteins from whey or from the casein fraction in order to limit allergenicity as much as possible residual product. On the hand, researchers, scientists and industrials keep searching for new potential milk alternatives including hydrolyzed milk formulae, plant-based formulae and other milks from different mammalian species such as goat, sheep, donkey, mare and camel milks [11, 12, 15, 16].

2. Characteristics of cow's milk proteins

Milk proteins represent an important nutritional source due to their high biological value and the presence of essential amino acids. They are also the source of various dairy products due to the important techno-functional properties of its proteins. Cow milk is a heterogeneous mixture of proteins with different structural and physicochemical properties. As milks from all mammalian species, cow milk proteins are divided according to their solubility into two fractions: caseins (insoluble in acidic conditions) and whey proteins (soluble proteins). Indeed, caseins precipitate at their isoelectric pH which is located at 4.6, while whey proteins remain soluble in this pH level [17].

2.1 Caseins

Caseins are phosphoproteins which represent the most abundant protein fraction in milk. They represent approximately 80% of the total milk protein.

Caseins consist of 4 proteins which differ in contents of phosphorus, concentration, amino acid composition, isoelectric point (pI) and molecular weight (MW): alpha S1, alpha S2, beta, and kappa (α_{S1}, α_{S2}, β and κ). The α and β caseins are calcium sensitive caseins as they precipitate at a calcium concentration at 30 mM while the κ-casein remains in solution under these conditions. The β-casein represent 39% of total caseins, followed by α_{S1}, α_{S2} and κ caseins which represent 38%, 10% and 13% of total amounts of caseins, respectively (**Figure 2**) [19].

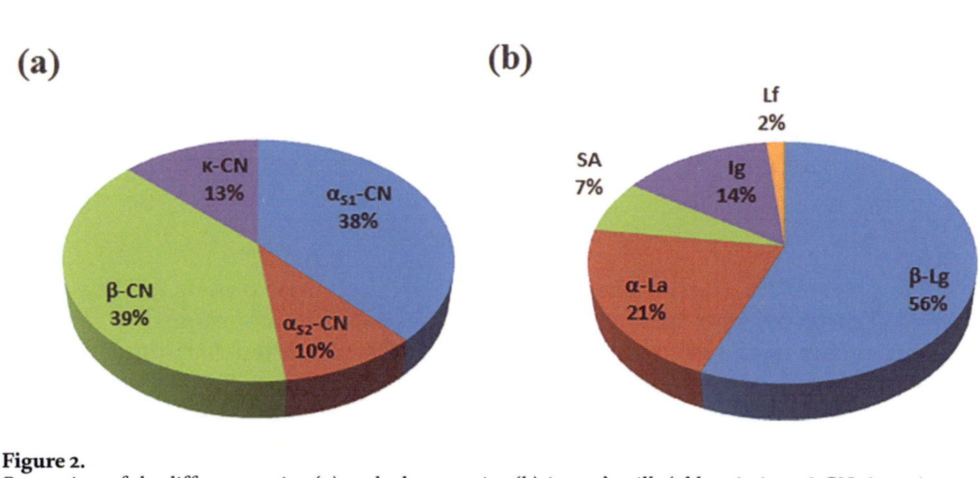

Figure 2.
Proportions of the different caseins (a) and whey proteins (b) in cow's milk (abbreviations: β-CN: β-casein; α_{S1}-CN: α_{S1}-casein; α_{S2}-CN: α_{S2}-casein; κ-CN: κ-casein, β-Lg: β-lactoglobulin; α-La: α-lactalbumin; SA: serum albumin; Ig: immunoglobulins; Lf: lactoferrin [18].

The 4 different caseins are associated with minerals forming colloids called casein micelles (concentration of minerals 80mg/g of caseins) with a diameter ranging between 100 and 140 nm. Bovine casein micelles and their characteristics have been the subject of much research and different micellar models have followed one another over the years (models by Horne, Holt, Bouchoux, etc.) [20, 21].

- **α_{S1}-Casein**: is a phosphoprotein with a MW of 22.9 kDa (199 amino acid residues) and is present in milk in the amount of 19.5 g/L and with pI of 4.46 [22]. Bovine α_{S1}-casein is characterized by the absence of cysteine residues in its molecular structure. Furthermore, no structural and functional homolog of animal α_{S1}-casein was observed in human milk. This is a major cause of the immunogenicity of this protein for humans and occurrence of CMA [23].

- **α_{S2}-Casein:** its concentration in milk is relatively low (3 g/L). It is composed of 207 amino acid residues and has a MW of 24.4 kDa and pI of 4.78. α_{S2}-Casein is the most hydrophilic of the caseins: it has 11 phosphorylated serine residues and is characterized by the presence of two cysteine residues (residues 36 and 40) creating intramolecular disulfide bridges. Hence, this casein is found in milk partly in dimeric form: two polypeptides which are linked by two disulfide bridges [24]. Four genetic variants were observed including variants A, B, C and D, while the A variant is the most common [23].

- **β-Casein**: is a phosphoprotein with a MW of 23.5 kDa, composed of 209 amino acid residues with a pI of 4.49. The concentration of this protein in cow milk is 11.7 g/L. 12 genetic variants were found for β-casein while the most common variants are A1, A2 and B. Homolog protein with similar structure and physico-chemical properties as bovine β-casein was found in human milk suggesting that this casein is the least allergenic casein in cow milk [23, 25].

- **κ-Casein:** is found in cow's milk at a concentration of 4.4 ± 0.3 g/L; thus, representing 13% of bovine caseins [26]. It is the least phosphorylated and the only glycosylated casein in milk from all mammalian species. The κ-casein has 169 amino acid residues with a MW of 18,974 kDa and a pI of 3.97. Furthermore, κ-casein has a particular amphipolar structure with a C-terminal which contains carbohydrate residues with a hydrophilic character and a hydrophobic N-terminal. It is also characterized by low calcium binding ability due to the presence of a single phosphorylation site at the residue 149. The κ-casein exhibits several biological functionalities such as anticoagulant properties as well as the prevention of platelet agglomeration and serotonin secretion [18].

2.2 Whey proteins

Soluble protein fraction or whey protein, is the second main protein fraction in milk (20–25% (w/w) of total protein). Overall, the protein composition of whey varies depending on the mammalian species. For cow's milk whey, the protein composition is as follows: β-lactoglobulin is the main protein (~56%), followed by α-lactalbumin (~21%), immunoglobulins (14%), bovine serum albumin (BSA) (7%) and lactoferrin (2%) (**Table 1**, **Figure 2**).

- **β-Lactoglobulin (β-Lg)** is a globular protein, present in the milk of all mammalian species except camelids, rodents and humans. The biological function of this

	Proteins	Allergen name	Molecular mass (kDa)	pI	Relative amount [a,b]	Amino acid residues	Allergenic activity (% of patients) [c]
Caseins, 80% (w/w) of total protein	α_{S1}-Casein	Bos d9	22.9	4.46	38%	199	57%
	α_{S2}-Casein	Bos d10	24.4	4.78	10%	207	
	β-Casein	Bos d11	23.5	4.49	39%	209	
	κ-Casein	Bos d12	18.9	3.97	13%	169	
Whey proteins, 20–25% (w/w) of total protein	β-Lg	Bos d5	18.28	5.2	56%	162	66%
	α-La	Bos d4	14.18	4.65	21%	123	18%
	BSA	Bos d6	66.4	4.7	7%	583	
	Lf	Bos d lactoferrin	76.1	8.18	2%	689	—
	Ig	Bos d7	150–800	5.5–7.5	14%	240–250 amino acids of the heavy chain	—

Abbreviations: pI: isoelectric point, α-La: α-lactalbumin, β-Lg: β-lactoglobulin, BSA: bovine serum albumin, Lf: lactoferrin, Ig: immunoglobulins.
[a] Proportion of individual caseins in the whole casein fraction of milk.
[b] Percentage of globular whey protein in the soluble fraction of milk.
[c] The allergenicity of the main proteins as reported by El-Agamy and Peñas et al. [13, 27].

Table 1.
Physico-chemical characteristics of the main cow's milk proteins and their allergenic activity (% of patients) [18].

protein is to transport the fatty acids, retinol and vitamins (A, D), binding Cu^{2+} and Fe^{2+} ions and inhibiting autooxidation of fats during digestion [28]. The β-Lg is the major protein in the soluble fraction of cow's milk with a concentration ranging between 2 and 4 g/L representing approximately 56% of the total bovine whey proteins [29]. The secondary structure of β-Lg consists of 10% α-helices, 45% β-sheets. It has two disulfide bonds at the cysteine residues (Cys106-Cys119 and Cys66-Cys160) and one free cysteine (Cys121) [30]. This protein is characterized by different quaternary structures depending on the environmental conditions of the protein (pH, temperature, ionic strength). The β-Lg, the primary structure comprises 162 amino acid residues with a MW of 18.281 kDa and a pI of 5.2. The β-Lg is an allergenic protein present due to its highest proportion among whey proteins (56% of total whey proteins) and due to the fact that this protein is totally absent in human milk [11, 31]. Food allergies associated with this allergenic protein may be present even in 80% of the total population [32].

- **α-Lactalbumin** (α-La): The α-La is a less allergenic protein than β-Lg and constitutes 21% of total whey protein [25]. Furthermore, chemical composition of bovine and human α-La bears a strong resemblance. This protein is a small protein of 123 amino acid residues (14.186 kDa, pI 4.65) known for its high content in the essential amino acids and for its important role in the biosynthesis of lactose with lactose synthetase and UDP galactosyl-transferase [33]. The α-La is a metalloprotein that contains one Ca^{2+} atom per mole of protein molecule, a divalent cation that plays an important role in stabilizing its spatial structure. The binding of this calcium ion is affected by the acid functions of the aspartic acid residues located in position 82, 87 and 88. There is also a second calcium binding site occupied by the zinc, but which has an affinity 105 times lower than that of calcium. The α-La contains 4 disulfide bridges (Cys6/Cys120, Cys28/Cys111, Cys61/Cys77 and Cys73/Cys91) but no free thiol groups. This configuration makes it more resistant to the phenomenon of protein aggregations caused by heat treatment, even though its denaturation temperature is relatively low (~64°C) [34].

The tertiary structure of this protein contains

- a β domain formed by β sheets. This domain has 10 Asp residues: it is acidic and represents the binding site of the Ca^{2+} ion, its pI is 3.37.

- an α domain consisting of four α helices forming a hydrophobic core. This domain is basic, containing 9 Lys residues with a pI 9.6

- **Lactoferrin** (Lf) is a protein synthesized by secretory epithelial cells of the mammary gland. It is a glycoprotein that belongs to the transferrin family containing two iron cation binding sites and more preferably the ferric ion (Fe^{3+}). This ability to scavenge for iron ions persists even at low pH values in the stomach and intestines, in order to deplete free iron which could slow bacterial growth in the intestine [35, 36]. The concentration of Lf in milk varies according to the producing animal species and according to the stage of lactation. The main function of this protein is binding iron and transporting it to the intestinal vascular system. Lf supports immune systems functionality, detoxification processes, as well as antineoplastic effect by inhibiting the attachment of tumor growth factors [37].

- **Bovine Serum Albumin** (BSA)—similarly, to caseins, β-Lg and α-La, this protein may also be a milk allergen. BSA is a whey protein characterized by its relatively high molecular mass. Indeed, bovine serum albumin (BSA) consists of 583 amino acids residues with a molecular mass of 66.4 kDa, its primary sequence has been determined by Hirayama et al. [38]. It has 17 intramolecular disulfide bridges and one free thiol group. This protein is present with w relative low concentration of 0.36 g/L in cow milk. This protein is inactivated at a temperature of 70–80°C. Among all cow's milk proteins probably only bovine serum albumin remains immunoreactive after heat treatment [25].

- **Lactoperoxidase** and **lysozyme** are active enzymes with antibiotic-like activity. For the lactoperoxidase, it is an oxidoreductase with antibacterial function, antineoplastic agent and viral growth inhibitor. On the other hand, lysozyme in milk has antiviral and anti-inflammatory properties.

3. Allergenicity of cow's milk proteins

Cow's milk contains approximately 30–35 g/L of proteins divided into 30 proteins, some of them are potentially allergenic and called "Bos d" and numbered according to the protein type [39].

The main cow milk allergens in are caseins (Bos d8) including β-casein (Bos d11), α_{S1}-casein (Bos d9), α_{S2}-casein (Bos d10) and κ-casein (Bos d12). On the other hand, whey consists of high allergenic proteins including α-La (Bos d 4), β-Lg (Bos d 5), immunoglobulins (Bos d7), BSA (Bos d6) and traces of Lf (Bos d Lf) [31, 40].

Scientists confirmed that the most commonly allergens which are usually detected in cow milk allergic patients are whole caseins especially the α_{S1}-casein (Bos d9), β-Lg (Bos d5) and α-La (Bos d4). Indeed, 66% of CMA is caused by the main cow milk allergen which is the β-Lg, followed by caseins (Bos d8) and significantly less by α-La and BSA (18%) [13, 25]. The high allergenicity of the β-Lg is attributed to the fact that this protein is totally deficient in human milk. Indeed, IgE response against β-Lg precedes those against the other allergens including caseins and α-La since birth. Afterwards, before the age of 1 year, IgE response toward caseins becomes predominant, whereas, the IgE response to α-La appears later after the age of 1 year [41].

However, the major problem of CMA is the fact that that only 27% of total patients with CMA are allergenic to only one allergen, meanwhile, the other patients present sensitization to two and more cow milk allergens leading to conclude that none of the main milk proteins allergens can be considered as the only responsible for the allergenicity of this food [11]. IgE do not react entirely with the antigenic protein but only with its allergenic part which is called epitope. Hence, one allergenic protein may have several epitopes, which might be the same or different depending on its quaternary structure and its exposed allergenic peptides. Epitopes of proteins molecules include immunodominant epitopes, which are the high allergenic epitopes and the main targets of immune response system. Allergy can not only be caused through bloodstream by the absorption of allergen but also by direct skin contact with the allergen [13].

The allergenicity of proteins, as well as the IgE epitopes of milk proteins, can be mapped and carried out using various bioinformatic tools through an *in silico* analysis. Overall, the primary protein sequences were taken from the UniProtKB protein Blast database, while the three-dimensional structures (downloadable as a pdb file) are listed in the PDB Protein Data base.

PD index, Bepipred, AlgPred, Discotope-2.0, Ellipro (prediction of linear and discontinuous epitopes) are some bioinformatic tools to study the allergenicity of proteins:

1. *Measuring the PD index (PD index)* using the physico-chemical properties of amino acids rather than their substitution frequencies in related proteins. Peptides or proteins with PD values less than 10 are considered to have significant physico-chemical similarities [42].

2. *BepiPred* predicts the location of linear B cell epitopes using a combination of a hidden Markov model and a propensity scale method. Peptides or proteins with a score greater that a value of 0.35 are suggested to be part of an epitope and stained yellow on the graph (where the Y axes represent residue scores and residue positions on the X axes in the sequence) [43].

3. *AlgPred* can be used for the prediction of the binding between the antigenic determinant and IgE. AlgPred can predict allergens by amino acid sequences by citing representative peptide sequences of allergens and this is based on the similarity of the known epitope to any region of protein. The SVM (Support vector

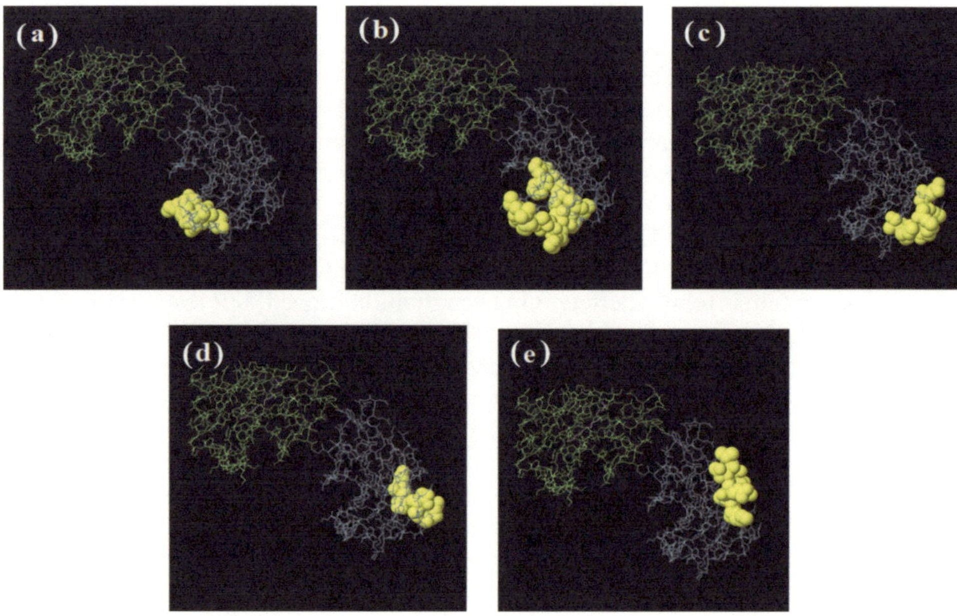

Figure 3.
An example of discontinuous B-cell epitopes predicted by the ElliPro. (a–e) Three-dimensional representation of conformational or discontinuous epitopes of bovine β-Lg. The epitopes are represented by yellow surface, and the bulk of the protein is represented in gray sticks.

machine) method of AlgPred calculates the allergenicity score of the protein that qualifies as "Allergen" for a score ≥ -0.5 [44].

4. *Discotope-2.0* is used for the prediction of discontinuous B cell epitopes from the 3D structure of the protein (pdb file). The method uses the calculation of surface accessibility and presented in terms of contact numbers. Final scores are calculated by combining the propensity scores of the spatial proximity residuals and the contact numbers [45].

5. *Ellipro* can also be used for the prediction of epitopes with a risk of cross-reaction between proteins. Ellipro can predict the protein's epitope based on peptides with a high allergenicity score as shown in **Figure 3**. Ellipro also makes it possible to present the potential epitopes on a 3D structure of the protein [46].

4. The effect of different processes on the allergenicity of cow's milk proteins

Food processing and additional ingredients cause changes in immunodominant epitopes and hence, the allergenic properties of proteins. Food processing may lead to the destruction of epitopes structures and/or the formation of new epitopes which are called neo-allergens. On the other hand, food processing can be associated with the reduction of allergenic properties of proteins or/and can have no influence on their allergenicity, it can even increase the immunogenicity of the treated proteins by the appearance of new epitopes [47].

Type of technological process	Operating conditions	The effect of process of the allergenicity of cow milk protein	The reference
Heat treatment	Pasteurization 90°C during 15 s	Low decrease on the immunoreactivity of whey proteins such as α-La and β-Lg	Wróblewska and Jędrychowski [50]
	Pasteurization 90°C during 15 min		
	Ultrasound at 52°C during 60°C	A significant reduction of the immunorectivity of α-La and β-Lg	
	Heat treatment at 80°C and 90°C during 30 min	The reduction of the allergenicity of immunoglobulins in milk	Jost et al. [51]
	Heat treatment at 120°C for 20 min	The reduction of the allergenicity of α-La by 25% compared to the native protein	Kleber and Hinrichs [52]
	Heat treatment of the cow milk proteins powder at 500°F (260°C) for 3 min	• 68% of children (n = 100, mean age, 7.5 years; range, 2.1–17.3 years) tolerated heated milk. • Smaller skin prick test wheals for heated milk-tolerant subjects • Lower milk-specific and casein specific IgE and lower IgE/IgG4 ratios to both of caseins and β-Lg (compared subjects with allergy to heated milk)	Nowak-Wegrzyn et al. [53]
High-pressure-processing	High-pressure treatment at 200–600 MPa and (temperature between 30°C and 68°C)	An increase of the antigenicity of the treated β-Lg in the WPI solution, sweet whey and skim milk	Kleber and Hinrichs [52]
	High-pressure treatment at 200 and 400 MPa	• The increase of the binding to β-Lg specific IgG from rabbit, • No effects on the IgE from allergic patients	Chicón et al. [54]
	High-pressure treatment at 600 MPa	The distribution of the structure of casein micelles and the decrease of the immunogenic capacity of milk proteins	Bogahawaththa et al. [55]
	High-pressure treatment at 400 MPa during 50 min	The loss of the allergenicity of the β-Lg hydrolysates with chymotrypsin by the absence of anaphylactic symptoms	López-Expósito et al. [56]

Type of technological process	Operating conditions	The effect of process of the allergenicity of cow milk protein	The reference
Enzymatic hydrolysis	Trypsin alone or in combination with both of chymotrypsin and pepsin	The reduction of the allergenicity of β-Lg without eliminating it	Bonomi et al. and Monaci et al. [57, 58]
	The combination of pepsin and α-chymotrypsin	The reduction of allergenicity by selective proteolysis of both α-La and β-Lg with a degree of hydrolysis of 1–20% and depending and incubation time	Monaci et al. [58]
	Trypsin	Only 4/10 patients (n = 10) had IgE antibodies to undigested β-Lg, while all 10 patients had IgE antibodies to β-Lg hydrolysates	Haddad et al. [59]
		The increase of the allergenicity of β-Lg: the derived peptides showed a specificity to bind human IgE by ELISA assays.	Selo et al. [60]
	Pepsin in the pH range 2–4	No differences were found in the antigenic properties of the hydrolysates of α-La, β-Lg, BSA and immunoglobulin G at pH 2 or 3, An enhancement of antigenicity of all proteins at pH 4 except β-Lg	Schmidt et al. [61]
	Corolase 7092	The increase of the antigenicity of proteins including BSA and immunoglobulin G	Ena et al. [62]
The lactic acid fermentation	*Lactococcus lactis* ssp. *lactis* 136	Immunoreactivity of raw milk: 0.10% for α-La and 3.36% for β-Lg	Wróblewska and Jędrychowski [50] and Miciński et al. [25]
	Lactobacillus casei 2	Immunoreactivity of raw milk: 0.56% for α-La and 2.18% for β-Lg	
	Lactobacillus acisophilus 67L	Immunoreactivity of raw milk: 0.09% for α-La and 1.46% for β-Lg	
	Lactobacillus delbruecki ssp. *bulgaricus* S11	Immunoreactivity of raw milk: 0.09% for α-La and 1.46% for β-Lg	

Abbreviations: α-La: α-lactalbumin, β-Lg: β-lactoglobulin, BSA: bovine serum albumin, WPI: whey protein isolate.

Table 2.
The effect of food processes (heating, high pressure, enzymatic hydrolysis and lactic acid fermentation) on the allergenicity of cow milk proteins.

4.1 The effect of heat treatments

Heating is an important process in the manufacturing of dairy products in order to obtain bacteriologically safe products leading to extend their shelf life. During the heating process, various structural modifications occur in the milk proteins depending on temperature, heating time, and heating exchanger. The structural and chemical changes in heating milk proteins such as denaturation, aggregation and "Maillard reaction" may have significant impacts on the antigenicity level of milk proteins [3, 48]. Among cow's milk proteins, caseins are the most heat stable proteins contrary to globular whey proteins which are sensitive to heat treatment and start to denaturate at temperatures above 60°C in the following order: BSA (denaturation temperature 94.9°C) < β-Lg (denaturation temperature 79.6°C) < α-La (denaturation temperature 70.5°C) [49].

Sterilization and pasteurization, which are the major categories of thermal processes have a significant impact on structural and functional properties of milk proteins leading to the increased, reduced or similar allergenicity. of allergenicity [50]. Wróblewska and Jędrychowski [50] noted that pasteurization of milk at 90°C resulted in a low decrease on the immunoreactivity of whey proteins such as α-La and β-Lg, while ultrasound treatments at 52°C during 60 min reduced greatly the immunorectivity of these proteins (**Table 2**). On the other hand, Jost et al. [51] reported that heating whey proteins at a temperature ranging between 80°C and 90°C during 30 minutes reduces the immunoglobulins contents as well as their immunogenicity.

Other researches carried out with bovine whey proteins confirmed that the antigenicity of β-Lg and α-La increases when heating temperature rose from 50 to 90°C because of the exposure of allergenic epitopes buried inside the native molecule due to the unfolding of conformational structure during heat denaturation. However, the antigenicity of these proteins decreased significantly above 90°C. Furthermore, the antigenicity of α-La decreased by 25% compared with its native state when it is treated at 120°C for 20 min [52, 63].

Other researches have evaluated whether children (n = 100) with CMA can tolerate extensively heated milk proteins and they found that approximately 68% tolerated the extensively heated milk. Furthermore, Heated milk-tolerant subjects showed significantly smaller skin prick test wheals, lower milk-specific and casein specific IgE as well as lower IgE/IgG4 ratios to both of caseins and β-Lg when compared subjects with allergy to heated milk [53]. Hence, some manufacturers use denatured whey proteins for the production of hypoallergenic infant formulae [25].

4.2 The effect of high-pressure processing

High-pressure processing is considered as a suitable nonthermal alternative method for milk pasteurization when it is in the range of 300–600 MPa [64]. High-pressure processing can even preserve the organoleptic and nutritional properties of the treated foods. However, this process can also alter structural and physico-chemical characteristics of proteins and result in their denaturation of native milk proteins. Indeed, high-pressure leads to the denaturation of whey proteins as the β-Lg and the changes of the casein micelles structures by their disassociation [55]. These changes may also influence the allergenicity of milk proteins. For instance, high-pressure treatment (200–600 MPa) at a temperature ranging between 30 and 68°C increased the antigenicity of β-Lg in the WPI (whey protein isolate) solution, sweet whey and skim milk [52]. Another study indicated that the high-pressure processing caused a severe whey protein denaturation especially the β-Lg and the minor whey proteins (Immunoglobulins) but no effect was observed for the α-La. Indeed, a high-pressure processing at 600 MPa induced the formation of large protein aggregates involving both of β-Lg and κ-casein through the thiol/disulphide interchange reactions. Furthermore, this treatment can disturb the structure of casein micelles leading the alteration of the immunogenic capacity of milk proteins diminished at 600 MPa [55]. Chicón et al. [54] found that the high pressure treatment of the pure β-Lg and whey protein isolate solution at 200 and 400 MPa resulted in an increase of the binding to β-Lg specific IgG from rabbit, without any effect on the IgE from allergic patients (**Table 2**). This behavior can be explained by the exposure of the buried epitopes in the unfolded protein molecules becoming more accessible for the antibodies.

Several researches focused on the effect of high-pressure on milk proteins hydrolysates. For instance, it was reported that a significant high degree of hydrolysis was levels obtained in high pressure (600 MPa), in comparison to atmospheric pressure depending upon the used enzyme. This behavior is attributed to the increased enzyme availability of immunogenic hydrophobic areas which, as a result, intensifies hydrolysis [25, 57].

On the other hand, hydrolysates obtained via the enzymatic treatment of main allergen in cow milk: β-Lg under high-pressure may result in a lower antigenicity and IgE binding ability [3, 57]. Indeed, the evaluated *in vivo* allergenicity of the β-Lg hydrolysates with chymotrypsin indicated that the tested hydrolysates with high-pressure treatment at 400 MPa during 50 min resulted in the loss of the allergenicity of the studied protein by the absence of anaphylactic symptoms. These results demonstrate the safety of hydrolysates produced under high-pressure conditions for manufacturing of novel milk formulae [56].

Other studies carried out with milk protein hydrolysates have also reported that the application of high-pressure treatment during enzymatic hydrolysis can significantly reduce the antigenicity of the treated proteins due to the increase of accessibility of the potentially immunogenic regions to the enzyme [3, 27, 54, 57].

4.3 The effect of enzymatic hydrolysis

Proteolysis have been usually considered as an efficient process to reduce allergenicity of milk proteins by destroying their allergenic epitopes [65]. The enzymatic hydrolysates were prepared with the use of digestive enzymes including pepsin, trypsin and chymotrypsin in order to imitate potential digestion processes and to reduce intestinal activity and the activity of enzymatic system in children [58]. However, the differences in the types of enzymes in this process as well as hydrolysis model and the hydrolysis degree may result in some discrepancies in the composition of the resulted peptide and a residual antigenicity of the hydrolysates as well as their taste [3]. Previous researches showed that the overall antigenicity of whey protein can be reduced by hydrolysis with trypsin alone. Caseins including α-casein and β-casein also show sensitivity to trypsin (unlike immunoglobulins and BSA). However, Nakamura et al. [66] noted that using many enzymes at the same hydrolysis process including papain, neutrase, alcalase and protease is more efficient in reducing the allergenicity of whey proteins when compared to those treated with a single enzyme. Thus, the hydrolysis of β-Lg by trypsin alone or in combination with chymotrypsin and pepsin.

It was proved that hydrolysis of β-Lg (Bos d5) by trypsin alone or in combination with both of chymotrypsin and pepsin reduces its allergenicity without eliminating it, while the combination of both of enzymatic hydrolysis and heat treatment was reported to reduce greatly the allergenicity of β-Lg [57, 58].

The combination of pepsin and α-chymotrypsin is considered as the most effective combination of enzymes used for the reduction of allergenicity and act by a selective proteolysis of both allergens α-La (Bos d4) and β-Lg (Bos d5) with a degree of hydrolysis of 1–20% and depending and incubation time [58].

An innovative technique of preparing hypoallergenic formulae for newborns involves the combination of hydrolysates and probiotics, which reduces allergic symptoms. Probiotics, including *Lactococcus lactis*, *Lactobacillus rhamnosus* and *Bifidobacterium lactis* significantly reduced the severity of atopic dermatitis in breast-fed infants after 2 months of treatment. Indeed, probiotics probably participate in mucosal degradation of macromolecules, leading to reduced allergenicity of milk proteins [25, 67–69].

Despite all these advantages of the hydrolysis of milk proteins for the reduction of their allergenicity, some researches confirmed the increase of the allergenicity of proteins by the exposure of new epitopes that appeared upon hydrolysis treatment. For instance, Haddad et al. [59] detected serum IgE from allergic patients using radioallergosorbent tests with a total tryptic hydrolysate of β-Lg (Bos d5) even when no IgE response was detected with the native protein of β-Lg (**Table 2**). Schmidt et al. [61] reported that no differences were found in the antigenic properties of the whey protein hydrolysates including α-La, β-Lg, BSA and bovine immunoglobulin G at pH 2 or 3, whereas, at pH 4 a further decrease in pepsin hydrolysis resulted in enhancement of antigenicity of all these proteins except the β-Lg. In the same way, the enzymatic proteolysis with Corolase 7092 was reported to increase the antigenicity of proteins including BSA and immunoglobulin G by exposing more antigenic sites during hydrolysis [62]. *In vitro* tests of Selo et al. [60] showed that that tryptic hydrolysis retained and even enhanced the allergenicity of β-Lg. In fact, the derived peptides showed a specificity to bind human IgE by ELISA assays. These authors also noted that numerous epitopes are widely scattered all along the β-LG molecule. They may be located in hydrophobic parts of the protein molecule, inaccessible for IgE antibodies in the native conformation of the protein but become bio-available after hydrolysis processes [60].

4.4 The effect of fermentation

The lactic acid fermentation process may have a potential influence on the allergenicity of cow milk proteins. Thus, researches were conducted with the use of several mesophilic and thermophilic bacterial strains which are already used in the production of fermented dairy products [25]. This process did not show a significant influence on the allergenic properties; indeed, the *in vitro* studies were not consistent with those *in vivo*.

Many studies have reported that Lactobacillus fermentation can induce degradation of milk allergens. For instance, lactic acid bacteria *Lactobacillus casei*, which are defined as probiotics [58]. In the same way, *Lactonacillus rhamnosus* GG has the ability to reduce phagocytosis which is stimulated by milk allergens by blocking receptors involved in phagocytosis on neutrophils and monocytes. It can even modify clinical symptoms in children with dermatitis and eczema [25, 70].

Clinical investigations have noted that dietary consumption of fermented foods, such as yogurt, can alleviate some of the symptoms of atopy and might also reduce the development of allergies through a mechanism of immune regulation.

The consumption of fermented milk cultures containing lactic acid bacteria can enhance the production of both Type I and Type II interferons at the systemic level [71]. However, changes of cow milk protein antigenicity and allergenicity depend on the species of lactic bacteria as well as the conditions of fermentation (**Table 2**). Lactic acid fermentation can reduce 90% of the antigenicity of the β-Lg in skim milk and 70% of this protein in sweet whey compared with untreated samples [52].

Finally, the reduction in antigenicity suggested that during the fermentation process with *Lactobacillus*, some epitopes of proteins were destroyed. These results are very useful for the preparation of new fermented milk products with reduced antigenic properties [3].

5. Conclusion

Cow's milk is a high nutritious food. However, it should be noticed that it contains many proteins which are considered as major food allergens leading to induce allergic reactions especially in infants.

The challenge for the food scientist, nutritionists and physicians is to resolve the problem of the CMA by searching new cow milk alternatives and/or new dairy processes that may reduce the allergenicity of cow milk proteins. Some processing technologies (heating, high pressure, enzymatic hydrolysis and lactic acid fermentation) can be used to effectively reduce the allergenicity of milk proteins by optimizing and controlling the processing conditions. However, attention should be paid during modification of milk proteins upon the used processes in order to prevent the appearance of some new epitopes during processing which are buried inside the native molecule. On the other hand, *in vitro* tests should be carried out to further detect the residual allergenicity of proteins and ensure the edible safety of milk products obtained by processing technologies. These strategies should provide valuable support for the development of the hypoallergenic milk formulae especially to infants.

Author details

Roua Lajnaf[1,2]*, Sawsan Feki[2], Hamadi Attia[1], Mohamed Ali Ayadi[1] and Hatem Masmoudi[2]

1 Alimentary Analysis Unit, National Engineering School of Sfax, Sfax, Tunisia

2 Immunology Department, Habib Bourguiba University Hospital, Sfax, Tunisia

*Address all correspondence to: roua_lajnaf@yahoo.fr

References

[1] Helm RM, Burks AW. Mechanisms of food allergy. Current Opinion in Immunology. 2000;**12**:647-653

[2] Halmerbauer G, Gartner C, Schierl M, Arshad H, Dean T, Koller DY, et al. Study on the prevention of allergy in children in Europe (SPACE): Allergic sensitization in children at 1 year of age in a controlled trial of allergen avoidance from birth. Pediatric Allergy and Immunology. 2002;**13**:47-54

[3] Bu G, Luo Y, Chen F, Liu K, Zhu T. Milk processing as a tool to reduce cow's milk allergenicity: a mini-review. Dairy Science & Technology. 2013;**93**:211-223

[4] Fritsché R. Role for technology in dairy allergy. Australian Journal of Dairy Technology. 2003;**58**:89

[5] Just J, Beaudouin E, Deschildre A, Renaudin J-M. Allergies Alimentaires: Nouveaux Concepts, Affections Actuelles, Perspectives Thérapeutiques. Issy-les-Moulineaux, France: Elsevier Masson; 2017

[6] Moen ØL, Opheim E, Trollvik A. Parents experiences raising a child with food allergy; A qualitative review. Journal of Pediatric Nursing. 2019;**46**:e52-e63

[7] Cianferoni A, Muraro A. Food-induced anaphylaxis. Immunology and Allergy Clinics. 2012;**32**:165-195

[8] Fiocchi A, Terracciano L, Bouygue GR, Veglia F, Sarratud T, Martelli A, et al. Incremental prognostic factors associated with cow's milk allergy outcomes in infant and child referrals: The Milan cow's milk allergy cohort study. Annals of Allergy, Asthma & Immunology. 2008;**101**:166-173

[9] Skripak JM, Matsui EC, Mudd K, Wood RA. The natural history of IgE-mediated cow's milk allergy. The Journal of Allergy and Clinical Immunology. 2007;**120**:1172-1177

[10] Pereira PC. Milk nutritional composition and its role in human health. Nutrition. 2014;**30**:619-627

[11] Hazebrouck S. Laits de chèvre, d'ânesse et de chamelle: Une alternative en cas d'allergie au lait de vache? Innovations Agronomiques. 2016;**52**:73-84

[12] Izadi A, Khedmat L, Mojtahedi SY. Nutritional and therapeutic perspectives of camel milk and its protein hydrolysates: A review on versatile biofunctional properties. Journal of Functional Foods. 2019;**60**:103441

[13] El-Agamy EI. The challenge of cow milk protein allergy. Small Ruminant Research. 2007;**68**:64-72

[14] Ekezie F-GC, Cheng J-H, Sun D-W. Effects of nonthermal food processing technologies on food allergens: A review of recent research advances. Trends in Food Science and Technology. 2018;**74**:12-25

[15] Nayak C, Ramachandra CT, Kumar G. A comprehensive review on composition of donkey milk in comparison to human, cow, buffalo, sheep, goat, camel and horse milk. Mysore Journal of Agricultural Sciences. 2020;**54**:42-50

[16] Businco L, Giampietro PG, Lucenti P, Lucaroni F, Pini C, Di Felice G, et al. Allergenicity of mare's milk in children with cow's milk allergy. The Journal of Allergy and Clinical Immunology. 2000;**105**:1031-1034

[17] Lajnaf R, Trigui I, Samet-Bali O, Attia H, Ayadi MA. Comparative study on emulsifying and physico-chemical properties of bovine and camel acid and sweet wheys. Journal of Food Engineering. 2019;**268**:109741

[18] Lajnaf R. Propriétés technofonctionnelles du lait de dromadaire [PhD thesis]. Montpellier, France: University of Montpellier; 2017

[19] Davies DT, Law AJR. The content and composition of protein in creamery milks in south-west Scotland. The Journal of Dairy Research. 1980;**47**:83-90

[20] Ali MZ, Robinson RK. Size distribution of casein micelles in camels' milk. The Journal of Dairy Research. 1985;**52**:303-307

[21] Attia H, Kherouatou N, Nasri M, Khorchani T. Characterization of the dromedary milk casein micelle and study of its changes during acidification. Le Lait. 2000;**80**:503-515

[22] El Agamy EI. Bioactive components in camel milk. In: Bioactive Components in Milk and Dairy Products. Athens, USA: Wiley-Blackwell Publishers, University of Georgia; 2009. p. 159

[23] Farrell HM Jr, Jimenez-Flores R, Bleck GT, Brown EM, Butler JE, Creamer LK, et al. Nomenclature of the proteins of cows' milk—Sixth revision. Journal of Dairy Science. 2004;**87**:1641-1674

[24] Kappeler S, Farah Z, Puhan Z. Sequence analysis of *Camelus dromedarius* milk caseins. The Journal of Dairy Research. 1998;**65**:209-222

[25] Miciński J, Kowalski IM, Zwierzchowski G, Szarek J, Pierożyński B, Zabłocka E. Characteristics of cow's milk proteins including allergenic properties and methods for its reduction. Polish Annals of Medicine. 2013;**20**:69-76

[26] Omar A, Harbourne N, Oruna-Concha MJ. Quantification of major camel milk proteins by capillary electrophoresis. International Dairy Journal. 2016;**58**:31-35

[27] Peñas E, Snel H, Floris R, Préstamo G, Gomez R. High pressure can reduce the antigenicity of bovine whey protein hydrolysates. International Dairy Journal. 2006;**16**:969-975

[28] Sawyer L, Kontopidis G. The core lipocalin, bovine β-lactoglobulin. Biochimica et Biophysica Acta (BBA)-Protein Structure and Molecular Enzymology. 2000;**1482**:136-148

[29] De Wit JN. Nutritional and functional characteristics of whey proteins in food products. Journal of Dairy Science. 1998;**81**:597-608

[30] Sawyer L, Brownlow S, Polikarpov I, Wu S-Y. β-Lactoglobulin: Structural studies, biological clues. International Dairy Journal. 1998;**8**:65-72

[31] D'Auria E, Mameli C, Piras C, Cococcioni L, Urbani A, Zuccotti GV, et al. Precision medicine in cow's milk allergy: Proteomics perspectives from allergens to patients. Journal of Proteomics. 2018;**188**:173-180

[32] Bernatowicz E, Reklewska B. Bioaktywne skladniki bialkowej frakcji mleka. Przegląd Hodowlany. 2003;**71**:1-10

[33] Permyakov EA, Berliner LJ. α-Lactalbumin: Structure and function. FEBS Letters. 2000;**473**:269-274

[34] McGuffey MK, Otter DE, van Zanten JH, Allen Foegeding E. Solubility

and aggregation of commercial α-lactalbumin at neutral pH. International Dairy Journal. 2007;**17**:1168-1178

[35] Campione E, Cosio T, Rosa L, Lanna C, Di Girolamo S, Gaziano R, et al. Lactoferrin as protective natural barrier of respiratory and intestinal mucosa against coronavirus infection and inflammation. International Journal of Molecular Sciences. 2020;**21**:4903

[36] Kappeler SR, Ackermann M, Farah Z, Puhan Z. Sequence analysis of camel (*Camelus dromedarius*) lactoferrin. International Dairy Journal. 1999;**9**:481-486

[37] Watrobinska K, Nalecz-Tarwacka T. Frakcje białkowe mleka krowiego. Przegląd Mleczarski. 2007;**6**:10-12

[38] Hirayama K, Akashi S, Furuya M, Fukuhara K. Rapid confirmation and revision of the primary structure of bovine serum albumin by ESIMS and Frit-FAB LC/MS. Biochemical and Biophysical Research Communications. 1990;**173**:639-646

[39] Apweiler R, Bairoch A, Wu CH, Barker WC, Boeckmann B, Ferro S, et al. UniProt: The universal protein knowledgebase. Nucleic Acids Research. 2004;**32**:D115-D119

[40] Hochwallner H, Schulmeister U, Swoboda I, Spitzauer S, Valenta R. Cow's milk allergy: From allergens to new forms of diagnosis, therapy and prevention. Methods. 2014;**66**:22-33

[41] Boutin A, Liabeuf V, Agabriel C, Cleach I, Vitte J. Profil de sensibilisation moléculaire aux protéines du lait de vache: Évolution de 0 à 16 ans. Revue Française d'Allergologie. 2015;**55**:219

[42] Naito M, Matsui T, Yamada C, Tagami K, Ito K, Izumi H. Evaluation of cross-reactivity between casein components using inhibition assay and *in silico* analysis. Pediatric Allergy and Immunology. 2021;**32**:544-551

[43] Jamakhani M, Lele SS, Rekadwad B. *In silico* assessment data of allergenicity and cross-reactivity of NP24 epitopes from *Solanum lycopersicum* (Tomato) fruit. Data Brief. 2018;**21**:660-674

[44] Saha S, Raghava GPS. AlgPred: Prediction of allergenic proteins and mapping of IgE epitopes. Nucleic Acids Research. 2006;**34**:W202-W209

[45] Kringelum JV, Lundegaard C, Lund O, Nielsen M. Reliable B cell epitope predictions: Impacts of method development and improved benchmarking. PLoS Computational Biology. 2012;**8**:e1002829

[46] Ponomarenko J, Bui H-H, Li W, Fusseder N, Bourne PE, Sette A, et al. ElliPro: A new structure-based tool for the prediction of antibody epitopes. BMC Bioinformatics. 2008;**9**:1-8

[47] Sathe SK, Teuber SS, Roux KH. Effects of food processing on the stability of food allergens. Biotechnology Advances. 2005;**23**:423-429

[48] Felfoul I, Jardin J, Gaucheron F, Attia H, Ayadi MA. Proteomic profiling of camel and cow milk proteins under heat treatment. Food Chemistry. 2017;**216**:161-169

[49] Felfoul I, Lopez C, Gaucheron F, Attia H, Ayadi MA. A laboratory investigation of cow and camel whey proteins deposition under different heat treatments. Food and Bioproducts Processing. 2015;**96**:256-263

[50] Wróblewska B, Jędrychowski L. Wpływ modyfikacji technologicznych na zmianę właściwości

immunoreaktywnych białek mleka krowiego. Alergia Astma Immunologia. 2003;**8**:157-164

[51] Jost R, Monti JC, Pahud JJ. Whey protein allergenicity and its reduction by technological means. Food Technology. 1987;**41**:118-121

[52] Kleber N, Maier S, Hinrichs J. Antigenic response of bovine β-lactoglobulin influenced by ultra-high pressure treatment and temperature. Innovative Food Science and Emerging Technologies. 2007;**8**:39-45

[53] Nowak-Wegrzyn A, Bloom KA, Sicherer SH, Shreffler WG, Noone S, Wanich N, et al. Tolerance to extensively heated milk in children with cow's milk allergy. The Journal of Allergy and Clinical Immunology. 2008;**122**:342-347

[54] Chicón R, Belloque J, Alonso E, López-Fandiño R. Immunoreactivity and digestibility of high-pressure-treated whey proteins. International Dairy Journal. 2008;**18**:367-376

[55] Bogahawaththa D, Buckow R, Chandrapala J, Vasiljevic T. Comparison between thermal pasteurization and high pressure processing of bovine skim milk in relation to denaturation and immunogenicity of native milk proteins. Innovative Food Science and Emerging Technologies. 2018;**47**:301-308

[56] López-Expósito I, Chicón R, Belloque J, López-Fandiño R, Berin MC. *In vivo* methods for testing allergenicity show that high hydrostatic pressure hydrolysates of β-lactoglobulin are immunologically inert. Journal of Dairy Science. 2012;**95**:541-548

[57] Bonomi F, Fiocchi A, Frøkiær H, Gaiaschi A, Iametti S, Poiesi C, et al. Reduction of immunoreactivity of bovine β-lactoglobulin upon combined physical and proteolytic treatment. The Journal of Dairy Research. 2003;**70**:51-59

[58] Monaci L, Tregoat V, van Hengel AJ, Anklam E. Milk allergens, their characteristics and their detection in food: A review. European Food Research and Technology. 2006; **223**:149-179

[59] Haddad ZH, Kalra V, Verma S. IgE antibodies to peptic and peptic-tryptic digests of betalactoglobulin: Significance in food hypersensitivity. Annals of Allergy. 1979;**42**:368-371

[60] Selo I, Clément G, Bernard H, Chatel J, Créminon C, Peltre G, et al. Allergy to bovine beta-lactoglobulin: Specificity of human IgE to tryptic peptides. Clinical & Experimental Allergy. 1999;**29**:1055-1063

[61] Schmidt DG, Meijer R, Slangen CJ, Van Beresteijn ECH. Raising the pH of the pepsin-catalysed hydrolysis of bovine whey proteins increases the antigenicity of the hydrolysates. Clinical and Experimental Allergy. 1995;**25**:1007-1017

[62] Ena JM, Van Beresteijn ECH, Robben A, Schmidt DG. Whey protein antigenicity reduction by fungal proteinases and a pepsin/pancreatin combination. Journal of Food Science. 1995;**60**:104-110

[63] Bu G, Luo Y, Zheng Z, Zheng H. Effect of heat treatment on the antigenicity of bovine α-lactalbumin and β-lactoglobulin in whey protein isolate. Food and Agricultural Immunology. 2009;**20**:195-206

[64] Trujillo AJ, Capellas M, Saldo J, Gervilla R, Guamis B. Applications of high-hydrostatic pressure on milk and dairy products: A review. Innovative Food Science and Emerging Technologies. 2002;**3**:295-307

[65] Heyman M. Evaluation of the impact of food technology on the allergenicity of cow's milk proteins. The Proceedings of the Nutrition Society. 1999;**58**:587-592

[66] Nakamura T, Sado H, Syukunobe Y, Hirota T. Antigenicity of whey protein hydrolysates prepared by combination of two proteases. Milchwissenschaft. 1993;**48**:667-670

[67] Adel-Patient K, Ah-Leung S, Créminon C, Nouaille S, Chatel J, Langella P, et al. Oral administration of recombinant Lactococcus lactis expressing bovine β-lactoglobulin partially prevents mice from sensitization. Clinical and Experimental Allergy. 2005;**35**:539-546

[68] Pessi T, Sütas Y, Marttinen A, Isolauri E. Probiotics reinforce mucosal degradation of antigens in rats: Implications for therapeutic use of probiotics. The Journal of Nutrition. 1998;**128**:2313-2318

[69] Isolauri E, Arvola T, Sütas Y, Moilanen E, Salminen S. Probiotics in the management of atopic eczema. Clinical and Experimental Allergy. 2000;**30**:1605-1610

[70] Wróblewska B. Białka pochodzenia zwierzęcego jako alergeny pokarmowe. Przemysł Spożywczy. 2007;**12**:61

[71] Cross ML, Stevenson LM, Gill HS. Anti-allergy properties of fermented foods: An important immunoregulatory mechanism of lactic acid bacteria? International Immunopharmacology. 2001;**1**:891-901

Section 5

Control Mechanism of Breast Milk

Chapter 6

Compare the Effects of Ultrasound versus Taping in Lactating Mothers with Breast Engorgement

Dasarapu Indrani, Jagatheesan Alagesan, Prathap Suganthirababu, M.V. Sowmya and Dubba NagaRaju

Abstract

Human milk has hundreds of milk proteins, which provides many benefits on breastfeeding. Breastfeeding is a mother's gift to herself, her baby, and the earth, there is no substitute for mother's milk. Milk protein is most important for baby's growth, development and protects the baby from different illness. Colostrum is produced during early days immediately after child birth, which contains important nutrients and antibodies. Breast engorgement is a problem that is commonly encountered in breastfeeding mothers, which is to be addressed and treated to provide good milk proteins to baby, by relieving discomforts of lactating mothers. A randomized controlled trial was conducted with 30 subjects based on inclusion and exclusion criteria where the subjects are divided into two groups, which contain 15 lactating mothers in each group. The control group that is group-A was treated with ultrasound, and the experimental group that is group-B was treated with ultrasound and Taping Technique. The result of the study showed that there was a significant difference between the pre- and posttest intervention, and we conclude that the ultrasound therapy and Kinesio taping was effective in treating lactating mothers with breast engorgement.

Keywords: milk protein, Lactating mothers, engorgement, VAS, SPES, ultrasound, taping, breastfeeding

1. Introduction

Milk protein is most important for baby's growth, development and protects the baby from different illness. Whey proteins and Casein proteins are two types of proteins in breast milk; whey proteins contain antibodies, lactoferrin and lysozyme, which protect baby from infection and are easy to digest. Casein proteins are harder to digest with more complex molecules. Colostrum is produced during early days immediately after child birth, which contains important nutrients and antibodies [1]. During early stages, lactating mothers may produce small quantity of colostrum; later milk production will be increased to the maximum, which makes the breast fuller and firmer causing increased blood flow and lymph fluids [2] to the breast tissue. If baby is not fed properly or any problem in lactating, the breast milk is stored, and some mothers may face problems related to over production of milk; all these components make the breast heavy and later turns to very hard as rock, this uncomfortable condition is known as breast engorgement. Breast engorgement problem [3] should be addressed because if it is left untreated, that may lead to serious issues [4], and in future it may result in painful blebs, and plugged milk ducts may also lead to mastitis. Without the knowledge of identification, many lactating mothers are suffering with breast engorgement [5]. Severe engorgement may also rise body temperature around 99–100 degree F, and this rise in body temperature is termed as "Milk fever." According to Academy of Breastfeeding Medicine Protocol Committee, breast engorgement is defined as "the swelling and distension of the breasts" [6]. Sometime engorgement results due to interrupted or infrequent or delayed milk from breast [7, 8]; this kind of problems may place the mother at high risk of engorgement [9], causing unhealthy growth and development of the infant. The infant may not get the required milk protein if the mother is facing problems of engorgement. This problem should definitely bring to the notice, which is troubling to both mother and infant where its incidence in the world is 1:8000 and in India is 1:6500. According to NFHS [10], painful breast problems are the most common reason for giving up breastfeeding. Treatment for breast engorgement can prevent future breast-related complication and also helps the baby to get proper milk proteins, which helps in proper growth and development of the baby. It is a major issue in early postpartum period as the breast under the influence of hormonal shift increased milk production rapidly. Interventions such as ultrasound therapy [11–14] application of hot moist [15], gentle massage [16] before feeding are beneficial. Through, Kinesio taping at the engorgement area, it decreases the inflammation, pain and improves circulation and lymphatic drainage. Hence, this study was to determine whether taping offers any advantage over ultrasound.

2. Methods

An experimental study was conducted on 30 subjects using convenient sampling technique based on inclusion and exclusion criteria. Lactating mothers with breast engorgement and pain for at least 2–3 days in postpartum period between 20 and 35 years of age were included in the study. Non-lactating women, pregnant women, lactating mothers with soft breast, lactating mothers receiving lactating suppressants, and lactating mothers with breast abscess, breast infection, mastitis, broken skin of the breast, bleeding or cracked nipple were excluded from the study. After receiving informed consent from the subjects, detailed explanation of the study is provided to them. All the information related to outcome measures, i.e., Six-Point Self-rated Engorgement Scale (SPES) [17] and Visual Analogue Scale (VAS), is given in **Figures 1** and **2**.

In this study, 30 subjects were divided into two groups, group-A and group-B. Fifteen subjects were allotted in each group. Pain parameter was measured with visual analogue scale in both groups before and after the treatment. Functional evaluation of both the groups was done with six-point self-rated engorgement scale before and after the treatment. Both the parameters were measured first day and after 1 week of treatment procedure. All the subjects received their treatment at the outpatient department of Saveetha College of Physiotherapy, Saveetha Institute of Medical and Technical Sciences, Chennai. Every subject followed the treatment for required period of 1 week.

Group-A: Fifteen subjects were treated with ultrasound for 1 week. The subjects were made to lie in supine with the arm of the treated side placed behind the head.

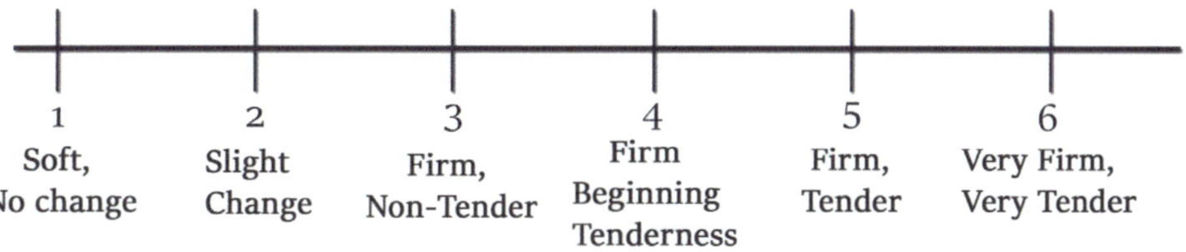

Figure 1.
Six-point self-rated engorgement scale.

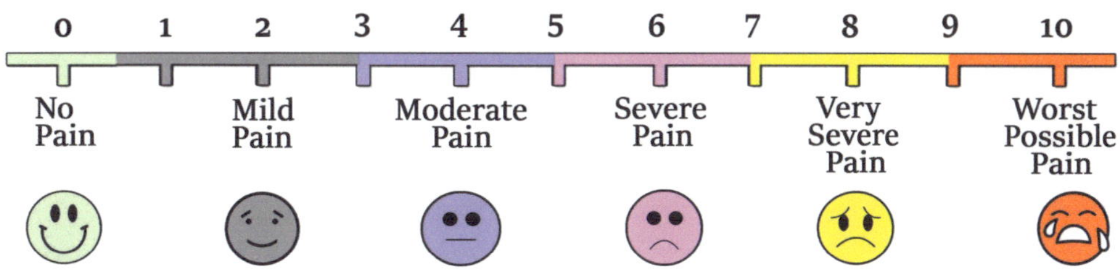

Figure 2.
Visual analogue scale.

A continuous mode of therapeutic ultrasound was given using ultrasound transmission gel as the coupling agent, with the intensity of 1 W/cm^2 and frequency of 1MHZ passing the head of the ultrasound firmly over the breast from the periphery toward the areola, lightly back to the chest, and firmly down again to the areola, gradually working around the breast for 8 minutes.

Group-B: To the next 15 patients, continuous mode of therapeutic ultrasound was given using ultrasound transmission gel as the coupling agent, with the intensity of 1 W cm^2 and frequency of 1MHZ passing the head of the ultrasound firmly over the breast from the periphery toward the areola, lightly back to the chest, and firmly down again to the areola, gradually working around the breast for 8 min, then subjects are treated with the taping techniques by using Kinesio tape (KT). Breast was exposed to clean with wet cotton dipped in water and breast is allowed to dry for few seconds and after drying, two pieces of tape which was about 7–9 inches, were taken. Seven to nine inches of tape was further cut into five strips equally. Taping was done with minimal stretch of 10–5% without extra tension by avoiding axilla with an anchoring base and rounded corners. Patients were instructed to wear the tape for 42–72 hours and also instructed to remove the tape prior to the prescribed time only if any skin irritation occurs. At the day 3 follow-up, the skin was inspected and assessed their primary outcome measures and then taped with the same technique used previously for 1 week. After the end of 1 week post Visual Analogue Scale and Six-Point self-rated Engorgement Scale were taken, and results are analyzed.

3. Results

Database was statistically analyzed using descriptive and inferential statistics; mean and standard deviation were estimated using paired and independent t test. Paired t test was used to compare data sets within the groups, and independent t test was used to compare the data sets between the groups (**Tables 1–4**).

Age distribution:

The average age of the subjects in group-A was 25.05 ± 2.04 years and in group-B was 25.25 ±.

1.82 years. There was no significant difference between the mean ages of the subjects in both the groups

Pre-test and post-test values of SPES and VAS of subjects in group A. The pre-test mean value of SPES was 4.53, and post-test mean value was 1.20. This shows that the

Groups	Mean Age(Yrs) + SD
Group-A	25.05 (± 2.04)
Group-B	25.25 (±1.82)

Table 1.
Mean age distribution.

GROUP-A		MEAN	STANDARD DEVIATION	t VALUE	P VALUE
SIX-POINT SELF-RATED ENGORGEMENT SCALE(SPES)	PRE TEST	4.53	1.06	11.3436	<0.0001
	POST TEST	1.20	0.41		
VAS SCORE (VAS)	PRE TEST	7.00	1.25	13.6871	<0.0001
	POST TEST	1.80	0.77		

Table 2.
Comparison of pre-test and post-test values of SPES and VAS in group-A.

(GROUP-B)		MEAN	STANDARD DEVIATION	t VALUE	P VALUE
SIX-POINT SELF-RATED ENGORGEMENT SCALE	PRE TEST	4.73	1.16	12.4335	<0.0001
	POST TEST	1.00	0.00		
VAS SCORE	PRE TEST	7.00	1.13	17.4611	<0.0001
	POST TEST	1.40	0.51		

Table 3.
Comparison of pre-test and post-test values of SPES and VAS in Group-B.

POST TEST		MEAN	STANDARD DEVIATION	t VALUE	P VALUE
SIX-POINT SELF-RATED ENGORGEMENT SCALE	Group A	1.20	0.41	1.8708	<0.0001
	Group B	1.00	1.16		
VAS SCORE	Group A	1.80	0.77	1.8708	<0.0001
	Group B	1.40	0.51		

Table 4.
Comparison of post-test values of SPES and VAS in groups A and B.

SPES was gradually decreasing significantly at $p < 0.0001$. The pre-test mean value of VAS was 7.0, and post-test mean value was 1.80.This shows that the VAS scores were gradually decreasing significantly at $p < 0.0001$.

Pre-test and post-test values of SPES and VAS of subjects in group-B. The pre-test mean value of SPES was 4.73, and post-test mean value was 1.00.This shows that the SPES scores were gradually decreasing significantly at $p < 0.0001$.The pre-test mean value of VAS was 7.00, and post-test mean value was 1.40.This shows that the VAS scores were gradually decreasing significantly at $p < 0.0001$.

Post-test values of SPES and VAS of subjects in group-A and group-B. The post-test mean value of SPES in group-A was 1.20, and post-test mean value of SPES in group-B was 1.00.This shows group-B has greater improvement in reduction of engorgement than group A with the p value (0.0001). The post-test mean value of VAS in group-A was 1.80, and post-test mean value of VAS of group-B was 1.40.This shows group-B has greater improvement in reduction of pain than group-A with the p value (0.0001).

Quantitative data analysis revealed that there is a significant difference between group A and B and within the groups. SPES post-test mean value in group-A was 1.20, and in group-B was 1.00. SPES Scores in group-B were comparatively lesser than those of group-A, $p < 0.0001$. The post-test mean value of VAS in group-A was 1.80, and post-test mean value of VAS in group-B was 1.40. This shows VAS scores in group-B were comparatively lesser than those in group-A, $p < 0.0001$.Statistical analysis of post-test for pain and engorgement revealed that subjects who received ultrasound and taping in group-B showed marked improvement compared with patients who received only ultrasound in group-A.

4. Discussion

Milk proteins are very essential for the baby in the early age of life; breast engorgement is a condition that troubles the baby as well as the mother by creating difficulties in breastfeeding, which is considered as second most problem affecting the lactation. To provide milk proteins and nutrients to the child, breast engorgement has to be treated in lactating mothers. So this study was designed to compare the effect of ultrasound and taping in lactating mothers with breast engorgement, with the help of ultrasound and taping technique, which reduces pain, engorgement and also prevents further complications. Reduction of engorgement helps the mother to feed her child and to provide proper milk proteins to the child. Breast milk is most important for the babies to get benefits of milk proteins. Breastfeeding plays an important role in reproductive age of women and beneficial for mother and child as well [18]. Breastfeeding is a physiological process, and it has to be encouraged; numerous studies demonstrate the importance of breastfeeding in providing protection against various diseases and decreasing the incidence of infant morbidity and mortality [19]. "All health professional groups support breastfeeding as the ideal way to nourish an infant, but numerous surveys have shown that, in general, even perinatal health professionals are not prepared to provide lactation management as part of routine care" [12]. Ultrasound helps the tissue to heal more effectively as it gives: 1) essential micromassage for individual cells, 2) increases cellular activity, and 3) responsible for the effect of therapeutic benefits. Ultrasound frequency was selected based on the depth of the tissue to be treated. The depth of ultrasound penetration was usually described in terms of half-value depth for the specific ultrasound frequency. Through, Kinesio taping at the engorgement area, it decreases the inflammation, pain and improves circulation and lymphatic drainage. In recent years, the use of Kinesio Tape (KT) has become increasingly popular. KT has same thickness as the epidermis in the skin when stretched to 30–40% of its resting length longitudinally, which is suitable to human skin, and was designed to mimic the qualities of human skin. It has roughly the same thickness as the epidermis and stretched between 30% and 40% of its resting length longitudinally. Kenzo Kase [20] proposed many benefits of Kinesio taping, which depends on the stretch applied during taping. It provides positional stimulus, creates sensory stimulation to limit motion, and removes edema. It is latex-free,heat-activated, and 100% cotton fiber helps to dry quickly. The purpose of our study was to investigate whether KT has an effect on breast engorgement in breastfeeding mothers during the postpartum period and also to help lactating mothers in providing proper milk proteins to the infant. We hypothesized that breastfeeding mothers would experience a decrease in breast engorgement by using the KT method, which helps in lactation and also provides milk proteins to the baby. Hence, the present study was undertaken with an intention to compare the effect of ultrasound therapy with Kinesio taping in lactating mothers. The result of the study showed that there was a significant difference between the pre- and post-test intervention.

5. Conclusion

In this study by comparing the effects of ultrasound versus taping in lactating mothers with breast engorgement, the result of the study showed that there was a significant difference between the pre- and post-test intervention. Both the groups resulted in positive outcomes, but group-B with ultrasound and Kinesio taping showed a higher level of positive outcome in terms of decreasing pain and engorgement, when compared with group-A with ultrasound among lactating mothers. The study concluded that ultrasound and Kinesio taping help in reducing pain and engorgement, which helps the mother to provide proper lactation, which in turn helps the baby to get proper milk proteins without delaying the feeding.

Author details

Dasarapu Indrani[1*], Jagatheesan Alagesan[2], Prathap Suganthirababu[2], M.V. Sowmya[2] and Dubba NagaRaju[3]

1 Urology and Obstetrics, Saveetha College of Physiotherapy, Saveetha Institute of Medical and Technical Sciences, Chennai, India

2 Saveetha College of Physiotherapy, Saveetha Institute of Medical and Technical Sciences, Chennai, India

3 NRI College of Physiotherapy, Andhra Pradesh, India

*Address all correspondence to: indranidasarapu@gmail.com

References

[1] Godhia M. Colostrums – its composition. Benefits as a Nutracentical-A Review. Current Research in Nutrition and Food Science. 2013;**1**(1):37-47

[2] Newton M, Newton NR. Postpartum engorgement of the breast. American Journal of Obstetrics and Gynecology. 1951;**61**(3):664-667

[3] Hill PD, Humenick SS. The occurrence of breast engorgement. Journal of Human Lactation. 1994;**10**(2):79-86

[4] Hewat RJ, Ellis DJ. A comparison of the effectiveness of two methods of nipple care. Birth. 1987;**14**(1):41-45

[5] Humenick SS, Hill PD, Anderson MA. Breast engorgement: Patterns and selected outcomes. Journal of Human Lactation. 1994;**10**(2):87-93

[6] Academy of Breastfeeding Medicine Protocol Committee, Eglash A. ABM clinical protocol#8:human milk storage information for home use for full-term infants (original protocol March 2004; revision#1 March 2010). Breastfeeding Medicine. 2010;**5**(3):127-130

[7] Lee WT, Lui SS, Chan V, Wong E, Lau J. A population-based survey on infant feeding practice (0-2 years) in Hong Kong: Breastfeeding rate and patterns among 3,161 infants below 6 months old. Asia Pacific Journal of Clinical Nutrition. 2006;**15**(3):377-387

[8] Priyanka P et al. Comparative effect of ultrasound therapy with conventional therapy on breast engorgement in immediate post-partum mothers. Integrative Molecular Medicine. 2016; **3**(2):553-558

[9] Arora S, Vatsa M, Dadhwal V. A comparison of cabbage leaves vs. hot and cold compresses in the treatment of breast engorgement. Indian Journal of Community Medicine. 2008;**33**(3):160

[10] Sagar K. Engorgement of breast-potential problem in lactation. Nightingale Nursing Times. 2004; **1**(5):17-21

[11] Mclachlan Z et al. Ultrasound threatment for breast engorgement: A randomized double blind trial. The Australian Journal of Physiotherapy. 1991;**37**(1):23-28

[12] Manna M, Devis PL. Effectiveness of hot fomentation versus cold compression on breast engorgement among postnatal mothers. International Journal of Nursing Research and Practice. 2016; **33**(3):160-123

[13] Mangesi L, Dowswell T. Treatments for breast engorgement during lactation (Review). The Cochrane Library. 2010;**9**:CD006946

[14] Newton M, Newton N. Postpartum engorgement of the breast. American Journal of Obstetrics & Gynecology. 1951;**61**:664-666

[15] Snowden HM et al. Treatment for breast engorgement during lactating. Cochrane Database Systematic Review. 2007;**2**:CD000046

[16] Öztürk G, Külcü DG, Mesci N, Şilte AD, Aydog E. Efficacy of kinesio tape application on pain and muscle strength in patients with myofascial pain syndrome: A placebo-controlled trial. Journal of Physical Therapy Science. 2016;**28**(4):1074-1079

[17] Del Ciampo LA, Del Ciampo IR. Breastfeeding and the Benefits of Lactation for Women's Health. Revista Brasileira de Ginecologia e Obstetrícia/RBGO Gynecology and Obstetrics. 2018;**40**(06):354-359

[18] Kacew S. Current issues in lactation: advantages, environment, silicone. Biomedical and environmental sciences: BES. 1994;**7**(4):307-319

[19] Naylor AJ, Creer AE, Woodward-Lopez G, Dixon S. Lactation management education for physicians. In Seminars in Perinatology. 1994;**18**(6):525-531

[20] Tantawy SA, Kamel DM. The effect of kinesio taping with exercise compared with exercise alone on pain, range of motion, and disability of the shoulder in postmastectomy females: A randomized control trial. Journal of Physical Therapy Science. 2016;**28**(12):3300-3305

Section 6

The Colostrum and Milk

Chapter 7

Colostrum and Milk in Sow

Morakot Nuntapaitoon

Abstract

Both colostrum and milk quality and quantity can influence piglet survival and growth, especially in a highly prolific sow. The Danish Landrace × Yorkshire crossbred was selected for high prolificacy and challenged to provide enough colostrum and milk of high quality to all piglets. This chapter reviewed the mechanism of colostrum and milk production, basic information of colostrum, and milk quality (immunoglobulin, fat, protein, lactose, etc.) and quantity. The importance of colostrum and milk in modern sows on piglet performance and survival was addressed. Since the sow immunoglobulin cannot pass epitheliochorial placenta in the sow to the piglet's bloodstream. Therefore, colostrum is a crucial role in piglet survival and growth. However, the amount of colostrum and milk production in hyperprolific sow still improve from high litter size. The knowledge about the factors influencing colostrum and milk quality and quantity, such as parity number, piglet, the environment in hyperprolific sows, may support veterinarians and farmers in the commercial swine farms for increasing pig production. Moreover, the technique to improve colostrum and milk quality and quantity were explained, such as feed supplementation in gestating and lactating sows.

Keywords: colostrum, milk, quality, quality, sow, how

1. Introduction

Colostrum and milk are important sources of energy in newborn and pre-weaning piglets. Milk secretion in sow was classified into three parts, including colostrum, transient milk and mature milk depending on composition in each period. Colostrum is the first secretion after farrowing until 24 hours after the first piglet is born [1]. Colostrum plays an important role in the survival and growth of piglets [2]. Colostrum composition includes nutrient, growth factors, hormone, and immune cells that relate to thermoregulation, growth, and contributing to intestinal development and glucose regulation. The transitional milk is a secretion from 34 hours after birth to day 4 of lactation, which was a high-fat concentration [3]. Mature milk is released during day 10 of lactation until weaning because milk composition is relatively stable [4]. Therefore, transition milk and mututre milk play in the piglet's growth and related both pre and post weaning performances. This chapter illustrates the importance, production, and composition of colostrum and milk in sow and their factors.

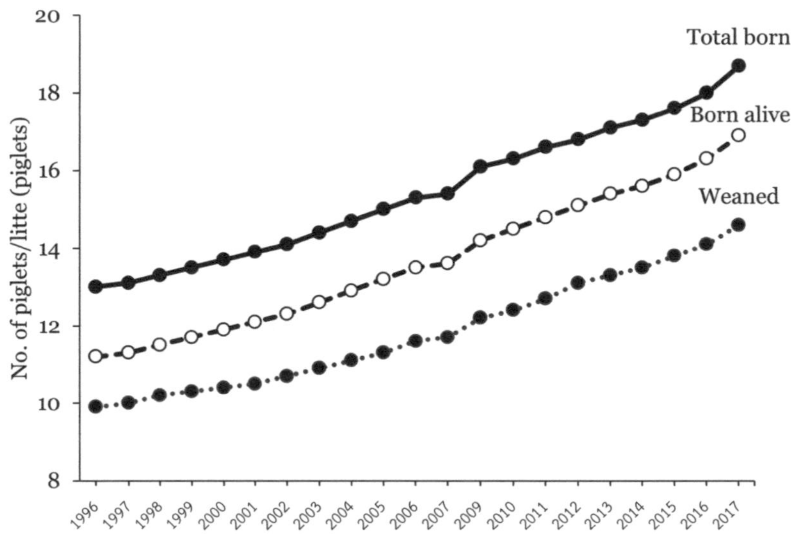

Figure 1.
Progress of the swine genetic improvement in Denmark during 1996–2017 total number of piglets per litter (Total number of piglets per litter: total born; number of born alive piglets per litter: Born alive; number of weaned piglets per litter: Weaned (modified from Observatori del porcí [6] and Hansen [7].

2. The important role of colostrum and milk

Colostrum and milk play an important role in piglet survival and growth during the lactation period. In the last decade, goal of genetic improvement in swine is to increase the total number of piglets per litter [5]. The total number of piglets per litter in Denmark in 1996–2017 rapidly increased from 13.0 to 18.7 piglet/litter or increasing 5.7 piglets/litter over the past 21 years (**Figure 1**). On the other hand, 50–80% of piglet mortality occurs during the first week after farrowing, especially in the first 72 hours of life [8–10]. In general, newborn piglets have glycogen storage in the liver and muscle for maintenance of the body, temperature, and energy for movement to consume colostrum after birth [11]. The glycogen rapidly declined within 12 to 17 h after birth when piglets have low colostrum consumption [12]. Therefore, it has been demonstrated that early mortality is mainly since a low colostrum consumption [13]. Furthermore, the relationship among colostrum consumption, mortality, and growth at weaning was reported in the previous study. Piglets consumed colostrum less than 400 g had a lower average daily gain than piglets consumed colostrum more than 400 g by 43 gram/day and higher mortality by 10 times [2].

Milk is a nutrient that most affected piglet growth during the suckling period. Milk supplementation in piglets improved growth performance that was reported in many studies [14, 15]. However, most of hyperprolific sows are low milk production, especially in tropical climates [4, 16]. Therefore, management to improve milk production in the lactation period should be concerned in commercial swine farms.

3. Mechanism of colostrum and milk production

Colostrum and milk were produced from the mammary gland of the sows that were developed from the embryo until entry to puberty and gestation. The mammary gland between birth and puberty was isometric growth and rapidly developing called

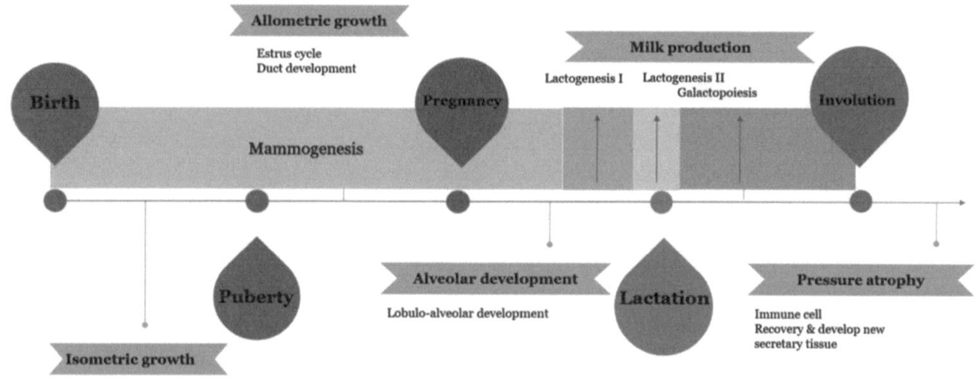

Figure 2.
Mammary gland development from birth until weaning.

"allometric growth" after the onset of puberty, gestation, and lactation (**Figure 2**). The mammary gland between puberty and pregnancy was provided by hormones for complete development, for example, growth hormone, prolactin, and estrogen.

In the mid of gestation, the mammary gland developed called "Lactogenesis I," which mainly developed duct and mammary gland by IGF-1 stimulation. The IGF-1 stimulates cortisol hormone from the adrenal gland and prolactin hormone from the placenta for inducing milk production. Completed alveolar development in the sows takes place during the last trimester of gestation called "Lactogenesis II" [17]. Prolactin induced lactoalbumin for producing lactose synthetase enzymes that were used for colostrum production. Colostrum was started to produced and kelp in the parenchyma tissue. Almost all colostrum is produced before the piglet is born and is independent of the suckling piglet activities [18]. However, the last week of gestation is crucial for colostrum production.

Most of the colostrum is secreted during the first 12 to 16 h after the onset of farrowing and decreases after 16 h onwards. Transient milk begins to produce during 24–34 h after the onset of farrowing within Lactogenesis II. The colostrum slowly changed to transient milk in this period. The stage of Lactogenesis II was finished within 1–2 days after farrowing. The transient milk slowly changed to mature milk on day 10 of lactation. After the colostrum period, milk secretion depended on the piglet's suckling activities to maintain milk secretion until weaning called "galactopoiesis." Galactokinesis or milk ejection is the active transfer of milk from the parenchyma to teats by suckling or other sensory activation (auditory, tactile, and visual). All activation stimulates oxytocin from the hypothalamus. Oxytocin is secreted into the blood and to the myoepithelial cell within the mammary gland leading to milk injection.

4. Calculation of colostrum and milk yields

The colostrum and milk yields represent the amount of colostrum and milk that were removed by piglets in the litter. Because yield was calculated from the sum of colostrum/milk intake of piglets. At present, there is no direct method to quantify both colostrum and milk yields.

Colostrum and milk yields measurement can be calculated from the indirect method, for example, the weigh-suckle-weigh method and predicted equations Devillers et al. [19]; Theil et al. [20, 21]; Hansen et al. [22]. See below.

4.1 Devillers et al.

Colostrum consumption (g) = $-217.4 + 0.217\,t + 1{,}861{,}019\,BW_{24}/t + BW_B (54.80-1{,}861{,}019/t)(0.9985-3.7{*}10^{-4}tfs + 6.1{*}10^{-7}\,tfs^2)$.

4.2 Theil et al.

Colostrum consumption (g) = $106 + 2{:}26\,WG + 200\,BW_B + 0{:}\,111\,D - 1414\,WG/D + 0.0182\,WG/\,BW_B$.

where t or D is time (min) elapsed between the 1st and 2nd weighting (which defines duration of colostrum consumption).

BW_{24} is body weight at 24 h (kg).

BW_B is birth weight (kg).

Tfs is the interval between birth and the first suckling (min).

WG is body weight gain between the 1st and 2nd weighting (g).

The predicted colostrum equation by Devillers et al. [19] was measured using bottle-fed-piglets but by Theil et al. [20] was measured using the deuterated water dilution technique. The previous study demonstrated that the predicted colostrum equation by Devillers et al. [19] was 43% lower than by Theil et al. [20] [3]. In line with this, according to the formula by Devillers et al. [19], a previous study demonstrated that piglets with the colostrum less than 200 g or 180 g/kg of birth weight have a high chance of mortality [1]. The piglets should be consuming 250 g of colostrum for survival and high growth performance, whereas Nuntapaitoon et al. [2] recommended that 200–400 g of colostrum should be provided in all piglets for decreasing mortality based on the formula by Theil et al. (**Figure 3**) [20].

Milk yield was also estimated by using the deuterated water dilution technique [21] and summarized data from many previous studies for generating predicted equation [22, 23]. For the latest equation, litter size and weight gain have to be included in the formula.

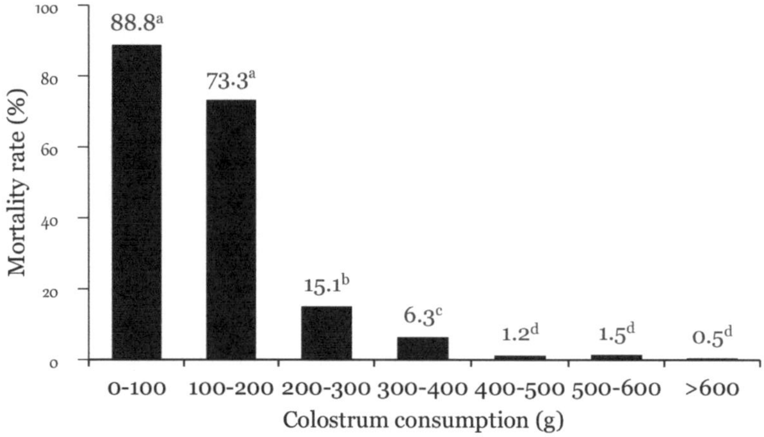

Figure 3.
Influence of colostrum consumption (g) on preweaning mortality in a commercial swine herd in a tropical climate calculated by Theil et al. [20]. Different superscript letters indicate significant differences ($P < 0.05$) *[2].*

5. Colostrum and milk yields

Amount of colostrum and milk yields were reported in many studies. The colostrum yield ranged 1.7–10.5 kg, and the colostrum consumption was 426 g piglet under tropical climate [2, 4, 24, 25]. The frequency distribution of individual colostrum consumption and colostrum yield in a commercial swine herd in Thailand was presented in **Figures 4** and **5**. On the other hand, range colostrum yield was 3.3–6.0 kg [4, 26–28].

Colostrum continuously releases during the colostral period. On the other hand, milk is released every 30–50 min and spends time 10–15 sec [29]. In general, milk yield in the first 4 days means 8 kg/day and the peak of lactation at 17 days was 15 kg/day [22, 23]. The milk yield ranged 3.9–17.2 kg/day in Danish Landrace ×Yorkshire crossbred sows reared in a commercial swine herd in Thailand. The frequency distribution of milk yield is presented in **Figure 6**. On the other hand, the range of milk yield was 5–15 kg/day [3, 30]. In line with this, colostrum and milk production in sows

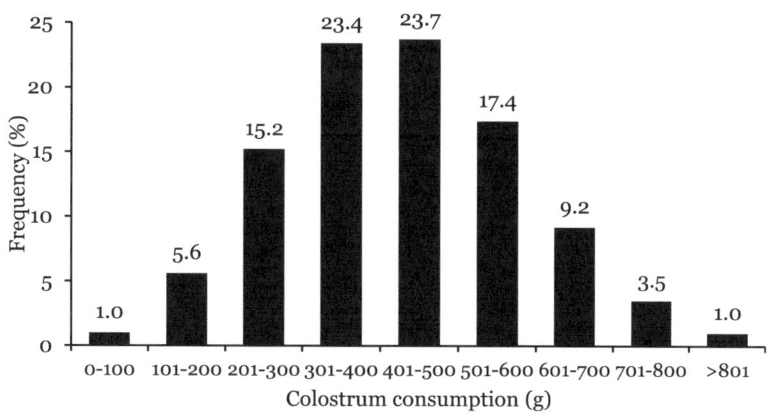

Figure 4.
Frequency distribution of individual colostrum consumption (g) in a commercial swine herd in Thailand [2].

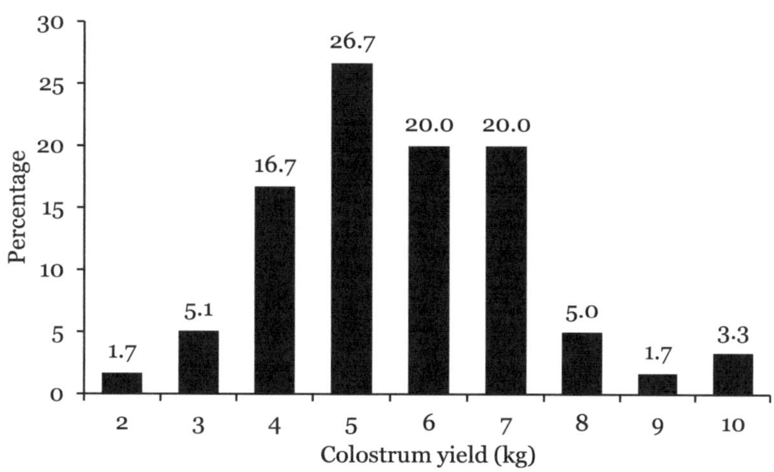

Figure 5.
Frequency distribution of individual colostrum yield (kg) in a commercial swine herd in Thailand [4].

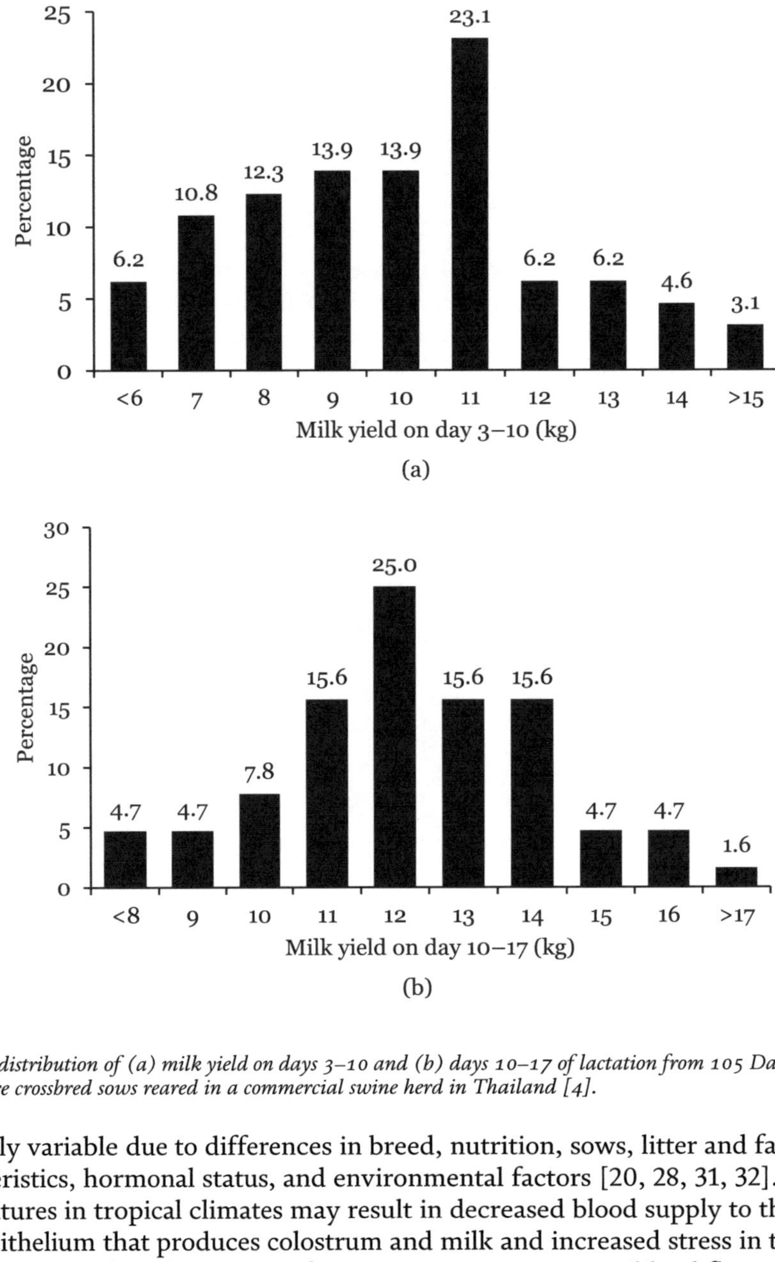

Figure 6.
Frequency distribution of (a) milk yield on days 3–10 and (b) days 10–17 of lactation from 105 Danish landrace × Yorkshire crossbred sows reared in a commercial swine herd in Thailand [4].

are highly variable due to differences in breed, nutrition, sows, litter and farrowing characteristics, hormonal status, and environmental factors [20, 28, 31, 32]. The high temperatures in tropical climates may result in decreased blood supply to the mammary epithelium that produces colostrum and milk and increased stress in the sows. Knowledge regarding the impact of temperature on mammary blood flow and colostrum and milk production is currently lacking.

6. Colostrum and milk composition

The main compositions of colostrum and milk include fat, protein, lactose, vitamin, mineral, and dry matter. Moreover, bioactive molecules, such as immunoglobulins, growth factors, and enzymes, are also included in milk secretion. It is important for the survival of the newborn piglet and the proper development of organs, such as

Composition	Days in milk							
	0		3		10		17	
	Mean	Range	Mean	Range	Mean	Range	Mean	Range
Lactose, g/100 g	2.8	1.7–5.0	4.3	2.1–5.3	4.8	2.1–5.5	4.9	2.4–5.6
Fat, g/100 g	4.9	2.5–7.5	7.1	4.0–14.2	6.1	2.3–10.9	6.1	3.9–8.7
Protein, g/100 g	15.3	5.2–22.3	7.2	3.7–16.7	5.7	3.8–16.7	5.9	3.7–18.5
DM, g/100 g	23.6	16.4–29.1	20.1	15.6–26.9	18.3	16.2–25.3	18.9	16.3–28.7

Nuntapaitoon et al. [4].

Table 1.
Colostrum and milk in Danish landrace × Yorkshire crossbred sows reared in a commercial swine herd in Thailand.

the gastrointestinal tract and brain. The main compositions of colostrum differ from milk (**Table 1**).

The chemical composition of colostrum (day 0), transition milk (day 3), and mature milk (day 10 and 17) is very different in hyperprolific sow. Lactose concentration gradually increases from 2.8 g/100 g in colostrum to 4.3 g/100 g in transition milk and 4.9 g/100 g in milk. Lactose is the main energy source for piglets throughout the lactation period. Moreover, lactose is rapidly absorbed and is related to the yield because of the structure and major osmotic characteristics of colostrum and milk [33].

Lipids were lowest in the colostrum period (4.9 g/100 g). It rapidly increases in the transition period (7.1 g/100 g) and is stable in mature milk (6.1 g/ 100 g). Fat also plays an important role to provide energy, increase metabolism, and protect the newborn against microbial infections. Many studies have analyzed fatty acids in various biological samples, such as plasma, milk, urine, and tissue samples, using a variety of analytical strategies. Analytical tools including gas-chromatography mass spectrometry (GC-MS), gas-chromatography with flame ionization detection (GC-FID), and liquid chromatography-mass spectrometry (LC-MS) have been used to perform fatty acid analyses [34]. The data from our team, a total of 31 free fatty acids in colostrum and milk of hyperprolific sows reared in a tropical climate, was presented in **Table 2** (unpublish data). It was found that free fatty acids in colostrum and milk are very different.

The most important in colostrum is immunity. Immunoglobulins as protein components in colostrum are important for piglets to prevent disease and reduce mortality. Colostrum contains six times immunoglobulins (IgA, IgG, and IgM) compared with milk. The concentration of IgG is rapidly declined by nearly 30% within 6 h after birth [35, 36] (**Figure 7**), while IgA slightly decrease. IgA is important for the protection of the gastrointestinal tract and plays a key role in preventing early diarrhea. The neonatal piglets have high morbidity from *Escherichia coli* and *Clostridium perfringens* infection [37] and lead to death in the first weeks of farrowing.

7. Factor influencing colostrum and milk yields and composition

Nutritional status in the gestation period highly influences colostrum and milk production in sows [3, 38]. Mammogenesis is started at 85–109 days of gestation [39].

Composition	Day		
	0	3	17
Fatty acid			
Caprylic acid	0.006[c]	0.012[b]	0.033[a]
Capric acid	0.062[b]	0.059[b]	0.207[a]
Lauric acid	1.2[b]	1.5[b]	3.1[a]
Myristic acid	4.5[b]	3.7[c]	5.8[a]
Myristoleic acid	0.025[c]	0.100[b]	0.287[a]
Pentadecylic acid	0.117	0.099	0.095
Palmitic acid	21.4[b]	21.2[b]	25.8[a]
Palmitoleic acid	1.7[a]	1.8[a]	0.6[b]
Margaric acid	0.262[a]	0.257[a]	0.200[b]
Cis-10-heptadecarnoic acid	0.115[c]	0.190[a]	0.157[b]
Stearic acid	6.5[b]	8.4[a]	5.9[b]
Elaidic acid	0.232[a]	0.170[b]	0.174[b]
Oleic acid	32.0[c]	37.6[a]	34.2[b]
Linolelaidic acid	0.145[a]	0.160[a]	0.101[b]
Linoleic acid	25.8[a]	19.0[b]	18.8[b]
Arachidic acid	0.157[b]	0.189[a]	0.194[a]
Gamma-Linoleic acid	0.239[b]	0.640[a]	0.192[b]
Paullinic	0.324[b]	0.634[a]	0.516[a]
Linolenic acid	1.8[a]	1.3[b]	1.4[b]
Eicosadienoic acid	0.565[a]	0.628[a]	0.402[b]
Behenic acid	0.049	0.047	0.060
Dihomo-ɣ-linolenic acid	0.316	0.272	0.189
Erucic acid	0.062[b]	0.083[a]	0.082[a]
Eicosatrienoic acid	0.176	0.114	0.088
Arachidonic acid	1.2[a]	1.0[b]	0.9[b]
Cis-13,16-docosadienoic acid	0.008	0.001	0.004
Lignoceric acid	0.179[a]	0.108[b]	0.097[b]
Docosatetraenoic acid	0.270[a]	0.205[ab]	0.187[b]
Docosapentaenoic acid	0.392	0.301	0.314
Eicosapentaenoic Acid	0.009	0.010	0.003
Docosahexaenoic acid	0.013	0.013	0.021

[a, b, c] *Different superscript letters within rows indicate significant differences* ($P < 0.05$).

Table 2.
The macrochemical composition and fatty acid profile in colostrum (day 0) and milk (day 3 and 17 of lactation).

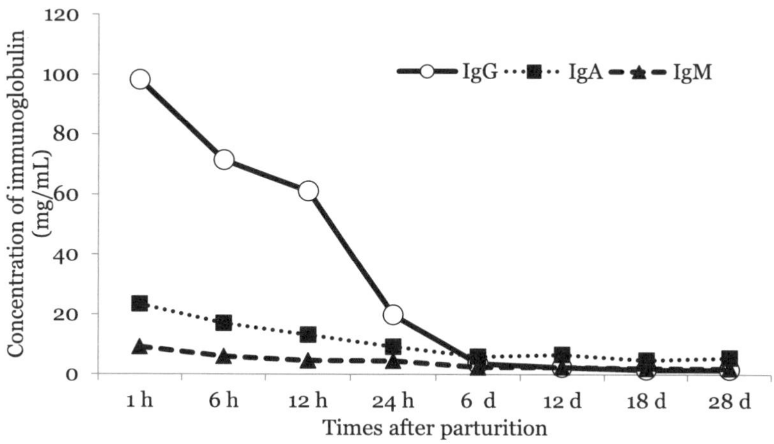

Figure 7.
The immunoglobulin concentration in sow colostrum throughout lactation period [35].

In this time, sow required more energy for developing mammary gland and also increase insulin resistance, especially in fat sows [40]. High insulin resistance presents high glucose level that passes through the mammary gland, leading to increased colostrum production. Many previous studies found that backfat thickness during late gestation influenced colostrum and milk production [24, 26]. They found that low-backfat thickness sows at 109 days of gestation had low milk production. The regression analyses revealed that an increase of backfat thickness by 1.0 mm at day 109 of gestation resulted in an increased milk yield of sows between 3 and 10 days of 271 g per day [24]. Recently, body weight at birth, cumulative birth interval, and litter size were significant risk factors affecting piglet colostrum consumption [41]. Furthermore, Nuntapaitoon et al. [4] found that sow parity number 2–4 had a higher colostrum yield than sow parity number 1 (**Table 3**).

The litter size increased milk production for stimulating the mammary gland by piglet [42]. **Figure 8** illustrated that high litter size is positively associated with milk production. However, high litter size declined individual colostrum consumption. In addition, piglet factors are also related to colostrum consumption [2, 41, 43]. They found that high piglet birth weight has high colostrum consumption and high suckling performance that stimulate milk production, especially in the first 3 days of lactation [44].

Sow parity number is the main association between production and composition. Multiparous sows have higher milk production than primiparous sows [4, 45]. The sow parity number 2–4 had the highest milk production and increased from first parity by 35% [45]. In contrast, Nuntapaitoon et al. [4] found that no evidence of parity differences was observed on milk yield.

Sow parity number has negatively correlated with fatty acid profiles in colostrum, which refers to metabolic status in sows [46]. The PLS-DA in **Figure 9a** shows the influence of parity number on the overall fatty acid profiles of colostrum. It has been demonstrated that significant dynamics in the fatty acid compositions of sow colostrum are in association with parity number. Moreover, high relative abundances of palmitic acid, eicosatrienoic acid, cis-10-heptadecanoic acid, capric acid, lignoceric

Composition	Parity			SEM*	*P*-value
	1	2–4	5–6		
Colostrum yield, kg	5.4[b]	7.0[a]	5.9[b]	0.4	0.004
Fat, g/100 g	5.2	4.9	4.6	0.2	0.164
Protein, g/100 g	15.4	15.2	15.2	0.9	0.979
Lactose, g/100 g	2.6	2.9	2.9	0.2	0.453
Dry matter, g/100 g	23.8	23.5	23.3	0.7	0.853

[a, b] *Different superscript letters within rows indicate significant differences (P < 0.05).*
**Greatest standard error of the mean (SEM).*
Nuntapaitoon et al. [4].

Table 3.
Effect of parity on colostrum yield and chemical composition of colostrum in Danish landrace × Yorkshire crossbred sows.

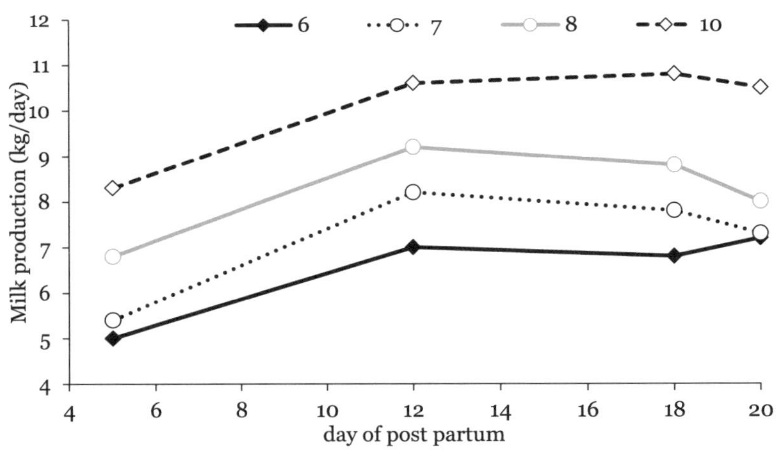

Figure 8.
Sow milk production in different litter sizes (modified by [42]).

acid, and lauric acid were accountable for the discrimination of colostrum from sows with higher parity numbers (**Figure 9b**). The high level of fatty acid profile in sow colostrum is related to the negative energy balance of sows. The stearic acid and palmitic acid have been related to negative energy balance periods, as animals mobilize adipose tissue for energy and related with colostrum production [47, 48], as in primiparous sows.

The concentration of immunoglobulin in colostrum depends on the management, the physiology of the mammary gland, parity, vaccination, and nutritional status [20, 28, 49]. In tropical climates, the variation of immunoglobulin concentration in the sow colostrum was influenced by their parity number and housing conditions [36]. The concentration of IgG in primiparous sows was lower than that in multiparous sows. Moreover, sow reared in a conventional open-housing system had a higher colostral IgG concentration than in an evaporative cooling-housing system. On the other hand, Zhao et al. [50] reported housing conditions did not relate to IgG concentration in colostrum. Therefore, factors influencing colostrum IgG concentration should be investigated in further study.

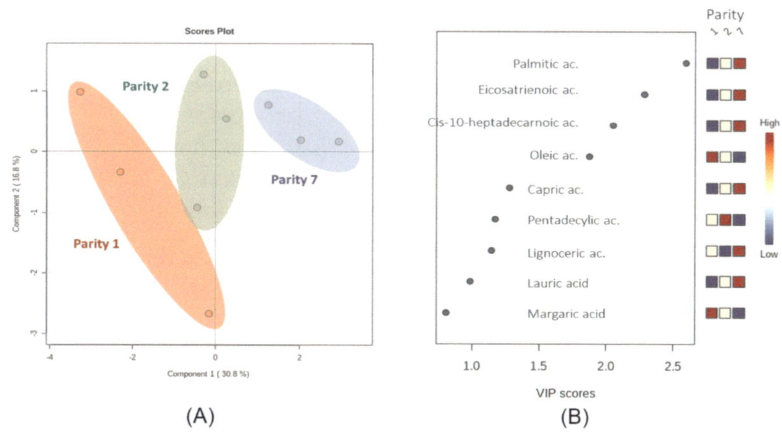

Figure 9.
PLS-DA score plot for an overall comparison of fatty acid profiles among representative colostrum samples from sows with parity number 1 (red), 2–6 (green), and ≥ 7 (blue) (panel A). VIP scores higher than 1.0 indicate potential biomarker fatty acids accountable for the discrimination among colostrum samples from different parity numbers (panel B) [46].

8. Technique for increasing colostrum and milk yield

The increasing feed intake and appetite in late gestating and lactating sow enhanced colostrum and milk production. Sow fed ad libitum in 7 days postpartum has higher milk production than in before farrowing [51]. Moreover, sow with appropriate condition before farrowing and peak of lactation related colostrum and milk production [18]. Therefore, many studies revealed the effect of feed additive and nutritional supplementation on colostrum and milk production many years ago. Protein supplementation in late-gestating sows improved colostrum production [52, 53]. They demonstrated that fermented potatoes protein increased colostrum yield, individual colostrum consumption in primiparous sows, and piglet birth weight and weight during the suckling in all sows period. It is illustrated that primiparous sows must be improved feed intake for colostrum production and fetal growth in the late gestation period. Increasing dietary protein at 135 g/day during lactation increased milk yield and milk protein concentration [54, 55].

Dietary fatty acid from different sources increases the amount of colostrum. Conjugated linoleic acids supplementation in late-gestating sow until farrowing increase +60 g of individual colostrum consumption [56]. Moreover, Flummer and Theil [57] found that supplementation of leucine increased colostrum consumption, increased growth rate, and decreased piglet mortality.

Fiber supplementation in late-gestating sow enhanced serum short-chain fatty acid in sow [22, 23]. The short-chain fatty acid is the source of milk production [18]. Quesnel et al. [58] reported that sows fed a high fiber diet from day 26 of gestation to farrowing had higher milk production than sows fed a low fiber diet. Loisel et al. [59] reported that fiber supplementation from 92 days of gestation to farrowing had increased colostrum production. However, Krogh et al. [60] compared different sources of fiber and fat during gestation on colostrum yield, that is, sugar beet pulp, alfalfa meal, and a combination of palm fatty acid distillate, soybean oil, and trioctanoate from day 105 of gestation. Different sources did not affect the colostrum yields of sow.

Generally, prostaglandin F2α was used for inducing farrowing in pregnancy sow and was also applied after parturition for reducing postpartum discharge and may affect colostrum and milk production. Milk synthesis collaborated hormones during parturition that declined progesterone and increased prolactin, estrogen, and corticosteroids. Luteolytic substance decreased serum progesterone concentration and increased prolactin, estrogen, and corticosteroids within 1 h after injection [33]. High concentration of progesterone declined milk synthesis [61]. Moreover, high progesterone levels in sow at the end of farrowing increase the risk of piglet diarrhea on the first day of life [62]. This is probably because declined colostrum production leads to low colostrum consumption and received low immunity. However, the previous studies demonstrated that farrowing induction did not affect colostrum production [25, 63–65] because colostrum is mostly produced before parturition at 85 days of gestation. On the other hand, recent research by Maneethong et al. [66] and Nuntapaitoon et al. [67] shows that natural prostaglandin F2α increased colostrum and milk production in the first week of lactation. The injected natural prostaglandin F2α after farrowing increased the milk yield between day 3 and 10 (**Figure 10**) [67]. In addition, prolonged farrowing duration declined colostrum yields [62]. In general, high litter size in hyperprolific sows increase farrowing duration. Piglet was born in prolonged farrowing time of sow highly chance hypoxia piglets and related colostrum consumption from high uterine contraction during peripartum period leading to decrease blood and oxygen supply to the piglets [4].

Furthermore, lactation management improved milk production. The sensory activation, such as auditory, also increased milk production and growth performance [68, 69]. The sow reared under temperature at 27–32°C has low milk production [70]. Farrowing pen easily assessed to mammary gland increased suckling behavior and milk production [71].

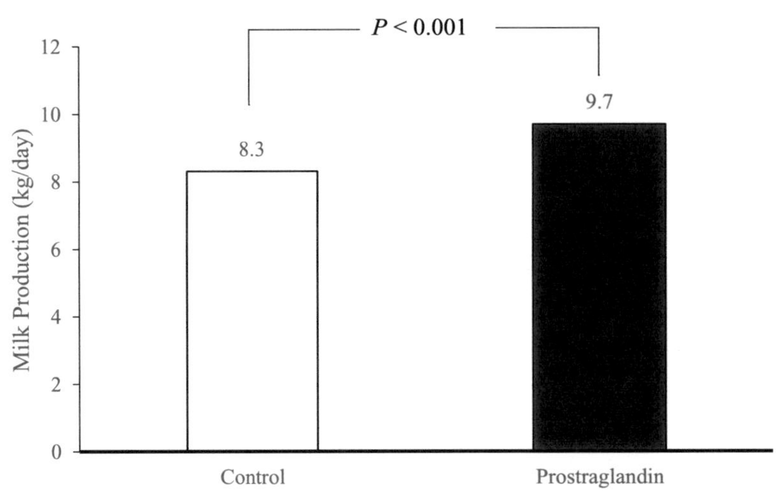

Figure 10.
The milk yield between days 3 and 10 in control and prostaglandin F2α group sows. A significant difference between a group at $P < 0.001$ *[67].*

9. Technique for increasing colostrum and milk quality

Increasing the fat content in the late gestational diet increase colostral fat. Kurachon et al. [72] found that protein supplementation in late-gestating sows increased colostral fat, especially in primiparous sows. Jackson et al. [73] reported that 10% corn oil supplementation during 100 days of gestation until farrowing increases colostrum fat. Moreover, Loisel et al. [59] and Krogh et al. [60] show that fiber supplementation in late gestation until farrowing increased the fat and lactose content of colostrum.

The effect of nutrition on the concentration of immunoglobulin in the colostrum was studied with particular attention to the increase in IgG intake. Dietary supplementation with conjugated linoleic acid in late gestation increased IgG, IgA, and IgM content in colostrum [74]. Moreover, Algae supplementation in sows at 107 days of gestation until weaning increased IgA concentrations and tended to increase IgG in colostrum. The Algae enhance protein and lysozyme in the sow and leading to increasing IgG concentrations in colostrum [75]. L-arginine supplementation in sow diet during late gestation increased immunoglobulin G concentration in colostrum. Nitric oxide synthase stimulates hormones and immune in sow that transfer to mammary tissue [76]. Dietary L-carnitine stimulated sow feed intake [77], and fat supplementation also enhanced milk fat and milk production [78]. Selenium plays an important role in colostrum and milk composition. The benefits of selenium improved versicular development in mammary tissue [79], immunoglobulin, and antioxidants in colostrum and milk [80, 81].

Moreover, many studies demonstrated that natural prostaglandin F2α increased IgG concentration in colostrum [66], and increased piglet survival and weaned weight [82–84] and was not negatively associated with sow reproductive performances [25, 64, 82]. However, Foisnet et al. [63] found that IgA concentration in colostrum declined when farrowing was induced by prostaglandin F2α. Recently, Taechamaeteekul et al. [65] illustrated that altrenogest in combination with double administrations of prostaglandin F2alpha did not affect colostral IgG. The benefits of prostaglandin F2α have not been clearly elucidated.

10. Conclusions

The quality and quantity of colostrum and milk are crucial for survival and growth in the piglets, especially in high prolific sows. The amount of colostrum and milk production represent sow health and performance in lactation. High immunoglobulin concentration transfers from sow still goals for protecting piglets. Fat and lactose in milk secretion are related to growth performance. The knowledge of improving colostrum and milk production and composition is still lacking. The nutritional strategies to increases piglet survival are the main further research. However, management in the late gestation thought out lactation period (i.e., induce farrowing, vaacination program and environment) also impacts piglet performances in both pre-and post-weaning periods.

Author details

Morakot Nuntapaitoon
Faculty of Veterinary Science, Department of Obstetrics, Gynaecology and Reproduction, Chulalongkorn University, Bangkok, Thailand

*Address all correspondence to: morakot.n@chula.ac.th

References

[1] Quesnel H, Farmer C, Devillers N. Colostrum intake: Influence on piglet performance and factors of variation. Livestock Science. 2012;**146**: 105-114

[2] Nuntapaitoon M, Muns R, Theil PK, Tummaruk P. Factors influencing colostrum consumption by piglets and their relationship with survival and growth in tropical climates. Livestock Science. 2019;**224**:31-39

[3] Quesnel H, Farmer C, Theil PK. Colostrum and milk production. In: The Gestating and Lactating Sow Gestating and Lactating Sows. Wageningen, The Netherland: Wageningen Academic Publishers; 2015. pp. 173-192

[4] Nuntapaitoon M, Juthamanee P, Theil PK, Tummaruk P. Impact of sow parity on yield and composition of colostrum and milk in Danish Landrace × Yorkshire crossbred sows. Preventive Veterinary Medicine. 2020;**181**:105085

[5] Marantidis A, Papadopoulos AI, Michailidisa G, Avdia M. Association of BF gene polymorphism with litter size in a commercial pig cross population. Animal Reproduction Science. 2013;**141**:75-79

[6] Observatori del porcí. Informe anual 2011. Departament d'Agricultura, Ramaderia, Pesca, Alimentació i Medi Natural. 2012. Available from: http://agricultura.gencat.cat/web/.content/de_departament/de02_estadistiques_observatoris/08_observatoris_sectorials/04_observatori_porci/informes _anuals/fitxers_estatics/op_inf_anual_2011.pdf

[7] Hansen C. National Average Productivity of Danish Pig Farms 2017. Report No. 1819. Copenhagen, Denmark: SEGES Danish Pig Research Centre; 2018. p. 19

[8] Koketsu Y, Takenobu S, Nakamura R. Preweaning mortality risks and recorded causes of death associated with production factors in swine breeding in Japan. The Journal of Veterinary Medical Science. 2006;**68**:821-826

[9] Shankar BP, Madhusudhan HS, Harish DB. Preweaning mortality in pig-causes and management. Veterinary World. 2009;**2**:236-239

[10] Wientjes JGM, Soede NM, Knol EF, van den Brand H, Kemp B. Piglet birth weight and litter uniformity: Effect of weaning-to-pregnancy interval and body condition change in sows of different parities and crossbred lines. Journal of Animal Science. 2013;**91**:2099-2107

[11] Amdi C, Krogh U, Flummer C, Oksbjerg N, Hansen CF, Theil PK. Intrauterine growth restricted piglets defined by their head shape ingest insufficient amounts of colostrum. Journal of Animal Science. 2013;**91**:5605-5613

[12] Theil PK, Cordero G, Henckel P, Puggaard L, Oksbjerg N, Sørensen MT. Effects of gestation and transition diets, piglet birth weight, and fasting time on depletion of glycogen pools in liver and 3 muscles of newborn piglets. Journal of Animal Science. 2011;**89**: 1805-1816

[13] Le Dividich J, Rooke JA, Herpin P. Nutritional and immunological importance of colostrum for the new-born pig. The Journal of Agricultural Science. 2005;**143**:469-485

[14] Harrell RJ, Thomas MJ, Boyd R. Limitations of sow milk yield on baby pig growth. In: Cornell Nutrition Conference, Syracuse New York, USA. UN: FAO; 1993. p. 55

[15] Miller YJ, Collins AM, Smits RJ, Thomson PC, Holyoake PK. Providing supplemental milk to piglets preweaning improves the growth but not survival of gilt progeny compared with sow progeny. Journal of Animal Science. 2012;**90**:5078-5085

[16] Khamtawee I, Singdamrong K, Tatanan P, Chongpaisarn P, Dumniem N, Pearodwong P, et al. Cinnamon oil supplementation of the lactation diet improves feed intake of multiparous sows and reduces pre-weaning piglet mortality in a tropical environment. Livestock Science. 2021; **251**:104657

[17] Senger PL. Pathways to Pregnancy and Parturition 3rd Edition. Pullman, United States: Current Conceptions Inc; 2015. p. 381

[18] Theil PK, Nielsen MO, Sorensen MT, Lauridsen C. In: Bach Knudsen KE, Kjeldsen NJ, Poulsen HD, Jensen BB, editors. Lactation, Milk and Suckling. In: Nutritional Physiology of Pigs. Copenhagen: Danish Pig Research Centre; 2012. pp. 1-47

[19] Devillers N, van Milgen J, Prunier A, Le Dividich J. Estimation of colostrum intake in the neonatal pig. Animal Science. 2004;**78**:305-313

[20] Theil PK, Flummer C, Hurley WL, Kristensen NB, Labouriau RL, Sørensen MT. Mechanistic model to predict colostrum intake based on deuterium oxide dilution technique data and impact of gestation and prefarrowing diets on piglet intake and sow yield of colostrum. Journal of Animal Science. 2014;**92**:5507-5519

[21] Theil PK, Kristensen NB, Jørgensen H, Labouriau RL, Jakobsen K. Milk intake and carbon dioxide production of piglets determined with the doubly labelled water technique. Animal. 2007; **1**:881-888

[22] Hansen AV, Strathe AB, Kebreab E, Theil PK. Predicting milk yield and composition in lactating sows-a Bayesian approach. Journal of Animal Science. 2012;**90**:2285-2298

[23] Hansen AV, Lauridsen C, Sørensen MT, Bach Knudsen KE, Theil PK. Effects of nutrient supply, plasma metabolites and nutritional status of sows during transition on performance in the following lactation. Journal of Animal Science. 2012;**90**:466-480

[24] Thongkhuy S, Chuaychu SHB, Burarnrak P, Ruangjoy P, Juthamanee P, Nuntapaitoon M, et al. Effect of backfat thickness during late gestation on farrowing duration, piglet birth weight, colostrum yield, milk yield and reproductive performance of sows. Livestock Science. 2020;**234**:103983

[25] Tospitakkul P, Kraomkaew K, Thammasin K, Uttarak P, Nuntapaitoon M, de Rensis F, et al. Induction of parturition by double administration of prostaglandin F2alpha in sows reduces the variation of gestation length without affecting the colostrum yield and piglet performance. The Journal of Veterinary Medical Science. 2019;**81**:1334-1340

[26] Decaluwé R, Maes D, Declerck I, Cools A, Wuyts B, De Smet S, et al. Changes in back fat thickness during late gestation predict colostrum yield in sows. Animal. 2013;7:1999-2007

[27] Devillers N, Farmer C, Le Dividich J, Prunier A. Variability of colostrum yield and colostrum intake in swine. Animal. 2007;**1**:1033-1041

[28] Quesnel H. Colostrum production by sows: Variability of colostrum yield and

immunoglobulin G concentrations. Animal. 2011;**5**:1546-1553

[29] Fisette K, Laforest J, Robert S, Farmer C. Use of recorded nursing grunts during lactation in two breeds of sows. I. Effect on nursing behavior and litter performance. Canadian Journal of Animal Science. 2004;**84**:573-579

[30] Zhou P, Nuntapaitoon M, Pedersen TF, Bruun TS, Fisker B, Theil PK. Digestibility and performance of lactating sows fed a coarsely ground diet. Journal of Animal Science. 2018;**96**:181-193

[31] Declerck I, Dewulf J, Piepers S, Decaluwé R, Maes D. Sow and litter factors influencing colostrum yield and nutritional composition. Journal of Animal Science. 2015;**93**:1309-1317

[32] Vadmand CN, Krogh U, Hasen CF, Theil PK. Impact of sow and litter characteristics on colostrum yield, time for onset of lactation, and milk yield of sows. Journal of Animal Science. 2015; **93**:2488-2500

[33] Foisnet A, Farmer C, David C, Quesnel H. Altrenogest treatment during late pregnancy did not reduce colostrum yield in primiparous sows. Journal of Animal Science. 2010;**88**:1684-1693

[34] Simionato JI, Garcia JC, dos Santos GT, Oliveira CC, Visentainer JV, de Souza NE. Validation of the determination of fatty acids in milk by gas chromatography. Journal of the Brazilian Chemical Society. 2010;**21**: 520-524

[35] Markowska ID, Małgorzata PM. Shifts in immunoglobulins levels in the porcine mammary secretions during whole lactation period. Bulletin-Veterinary Institute in Pulawy. 2010;**54**: 345-349

[36] Nuntapaitoon M, Suwimonteerabutr J, Am-in N, Tienthai P, Chuesiri P, Kedkovid R, et al. Impact of parity and housing conditions on concentration of immunoglobulin G in sow colostrum. Tropical Animal Health and Production. 2019;**51**:1239-1246

[37] Johansen M. Diarrhea in Newborn Piglets Caused by E-Coli Infections, Danish 'Colidiarré Hos spædgrise', Report. Axelborg, Copenhagen: The Danish Pig Research Centre; 2013

[38] Strathe A, Bruun T, Hansen C. Sows with high milk production had both a high feed intake and high body mobilization. Animal. 2017;**11**:1-9

[39] Ji F, Hurley WL, Kim SW. Characterization of mammary gland development in pregnant gilts. Journal of Animal Science. 2006;**84**:579-587

[40] Père MC, Etienne M, Dourmad JY. Adaptations of glucose metabolism in multiparous sows: Effects of pregnancy and feeding level. Journal of Animal Science. 2000;**78**:2933-2941

[41] Juthamanee P, Tummaruk P. Factors associated with colostrum consumption in neonatal piglets. Livestock Science. 2021;**251**:104630

[42] Toner MS, King RH, Dunshea FR, Dove H, Atwood CS. The effect of exogenous somatotropin on lactation performance of first-litter sows. Journal of Animal Science. 1996;**74**:167-172

[43] King RH, Mullan BP, Dunshea FR, Dove H. The influence of piglet body weight on milk production of sows. Livestock Production Science. 1997;**47**: 169-174

[44] Algers B, Jensen P. Teat stimulation and milk producion during early lactation in sows: Effect of continuous

noise. Canadian Journal of Animal Science. 1991;**71**:51-60

[45] Ngo TT, Quiniou N, Heugebaert S, Paboeuf F, Dourmad JY. Influence du rang de portée et du nombre de porcelets allaités sur la production laitière des truies. Journées Recherche Porcine. 2012; **44**:195-196

[46] Homyog K, Settachaimongkon S, Nuntapaitoon M. Correlation between sow parity number and fatty acid profiles in colostrum. In: Proceeding of the Chulalongkorn University Veterinary Conference (CUVC2021 live). Thailand: Chulalongkorn University Press; 2021. Thai J. vet. Med. 51 (Suppl.): 242-243

[47] Enjalbert F, Nicot MC, Bayourthe C, Moncoulon R. Duodenal infusions of palmitic, stearic or oleic acids differently affect mammary gland metabolism of fatty acids in lactating dairy cows. The Journal of Nutrition. 1998;**128**:1525-1532

[48] Mele M, Macciotta NPP, Cecchinato A, Conte G, Schiavon S, Bittante G. Multivariate factor analysis of detailed milk fatty acid profile: Effects of dairy system, feeding, herd, parity, and stage of lactation. Journal of Dairy Science. 2016;**99**:9820-9833

[49] Kirkden RD, Broom DM, Andersen IL. Piglet mortality: Management solutions. Journal of Animal Science. 2013;**91**:3361-3389

[50] Zhao Y, Flowers WL, Saraiva A, Yeum KJ, Kim SW. Effect of social ranks and gestation housing systems on oxidative stress status, reproductive performance, and immune status of sows. Journal of Animal Science. 2013; **91**:5848-5858

[51] Danielsen V. Fodringsstrategier for diegivende søer (Feeding strategies for lactating sows). In Danish. Vol. 33. Grøn Viden, Husdyrbrug; 2003. pp. 1-8

[52] Kurachon K, Jatutain P, Pongkhualao P, Phulakorn P, Ta-ngam T, Suwimonteeraburt P, et al. Dietary fermented potato protein supplementation in late gestating sows related colostrum production and piglet pre-weaning mortality. In: Proceeding of the Chulalongkorn university veterinary conference (CUVC2020). Nonthaburi, Thailand; 2020. Thai J. vet. Med. 50 (Suppl.): 265-266

[53] Kurachon K, Jatutain P, Pongkhualao P, Phulakorn P, Ta-ngam T, Suwimonteeraburt J, et al. Enhancing colostrum consumption and piglet performance in suckling period using fermented potatoes protein supplementation in late gestating sows. In: Proceeding of the International Pig Veterinary Society Congress (IPVS2020). Brazil: Rio De Janeiro; 2020. p. 625

[54] Strathe AV, Bruun TS, Tauson AH, Theil PK, Hansen CF. Increased dietary protein for lactating sows affects body composition, blood metabolites and milk production. Animal. 2020;**14**:285-294

[55] Strathe TS, Bruun TS, Geertsen N, Zerrahn JE, Hansen CF. Increased dietary protein levels during lactation improved sow and litter performance. Animal Feed Science and Technology. 2017;**232**:169-181

[56] Krogh U, Flummer C, Jensen SK, Theil PK. Colostrum and milk production of sows is affected by dietary conjugated linoleic acid. Journal of Animal Science. 2012;**90**:366-368

[57] Flummer C, Theil PK. Effect of β-hydroxy β-methyl butyrate supplementation of sows in late gestation and lactation on sow production of colostrum and milk and

piglet performance. Journal of Animal Science. 2012;**90**:372-374

[58] Quesnel H, Meunier-Salaün MC, Hamard A, Guillemet R, Etienne M, Farmer C, et al. Dietary fiber for pregnant sows: Influence on sow physiology and performance during lactation. Journal of Animal Science. 2009;**87**:532-543

[59] Loisel F, Farmer C, Ramaekers P, Quesnel H. Effect of high fibre intake during late pregnancy on sow physiology, colostrum production and piglet performance. Journal of Animal Science. 2013;**91**:5269-5279

[60] Krogh U, Bruun TS, Amdi C, Flummer C, Poulsen J, Theil PK. Colostrum production in sows fed different sources of fiber and fat during late gestation. Canadian Journal of Animal Science. 2015;**950**:211-223

[61] De Passillé AM, Rushen J, Foxcroft GR, Aherne FX, Schaefer A. Performance of young pigs: Relationships with periparturient progesterone, prolactin, and insulin of sows. Journal of Animal Science. 1993;**71**:179-184

[62] Hasan S, Orro T, Valros A, Junnikkala S, Peltoniemi O. Factors affecting sow colostrum yield and composition, and their impact on piglet growth and health. Livestock Science. 2019;**227**:60-67

[63] Foisnet A, Farmer C, David D, Quesnel H. Relationships between colostrum production by primiparous sows and sow physiology around parturition. Journal of Animal Science. 2011;**88**:1672-1683

[64] Otto MA, Machado AP, Moreira LP, Bernadi ML, Coutinho ML, VazJr IS, et al. Colostrum yield and litter performance in multiparous sows subjected to farrowing induction. Reproduction in Domestic Animals. 2017;**52**:749-755

[65] Taechamaeteekul P, Dumniem N, Sang-Gassanee K, Tummaruk P. Control of parturition in hyperprolific sows by using altrenogest and double administrations of PGF2α. Theriogenology. 2022;**181**:24-33

[66] Maneetong P, Srisang C, Sunanta N, Muchalintamolee P, Pearodwong P, Suwimonteerabutr J, et al. Postpartum prostaglandin F2α administration affects colostrum yield, immunoglobulin G, and piglet performance. Animal Bioscience. 2021;**34**:833-843

[67] Nuntapaitoon M, Arammanupunyakul J, Chatrapirat S, Boonraungrod N, Nimmansamai I, Tummaruk P. Using natural prostaglandin F2α before and after farrowing enhanced milk production in sows. In: Proceeding of the Chulalongkorn university veterinary conference (CUVC2021 Live). Thailand; 2021. Thai J. vet. Med. 51 (Suppl.): 299-300

[68] Cronin GM, Leeson E, Cronin JG, Barnett JL. The effect of broadcasting sow suckling grunts in the lactation shed on piglet growth. Asian-Australasian Journal of Animal Sciences. 2001;**14**:1019-1023

[69] Wechsler B, Brodmann N. The synchronization of nursing bouts in group-housed sows. Applied Animal Behavioral Science. 1996;**47**:191-199

[70] Farmer C, Prunier A. High ambient temperatures: How they affect sow lactation performance. Pig News and Information. 2002;**23**:95N-102N

[71] Pedersen ML, Moustsen VA, Nielsen MBF, Kristensen AR. Improved udder access prolongs duration of milk letdown

and increases piglet weight gain. Livestock Science. 2011;**140**:253-261

[72] Kurachon K, Jatutain P, Pongkhualao P, Phulakorn P, Ta-ngam T, Suwimonteeraburt J, et al. Improving colostrum composition using dietary fermented potato protein supplementation in late gestating sows. In: Proceeding of the International Pig Veterinary Society Congress (IPVS2020). Brazil: Rio De Janeiro; 2020. p. 626

[73] Jackson JR, Hurley WL, Easter RA, Jensen AH, Odle J. Effects of induced or delayed parturition and supplemental dietary fat on colostrum and milk composition in sows. Journal of Animal Science. 1995;**73**:1906-1913

[74] Corino C, Pastorelli G, Rosi F, Bontempo V, Rossi R. Effect of dietary conjugated linoleic acid supplementation in sows on performance and immunoglobulin concentration in piglets. Journal of Animal Science. 2009; **87**:2299-2305

[75] Leonard SG, Sweeney T, Bahar B, O'Doherty JV. Effect of maternalseaweed extract supplementation on suckling piglet growth, humoral immunity,selected microflora, and immune response after an ex vivo lipopolysaccharidechallenge. Journal of Animal Science. 2012;**90**:505-514

[76] Nuntapaitoon M, Muns R, Theil PK, Tummaruk P. L-arginine supplementation in sow diet during late gestation decrease stillborn piglet, increase piglet birth weight and increase immunoglobulin G concentration in colostrum. Theriogenology. 2018;**121**: 27-34

[77] Lauridsen C, Danielsen V. Lactational dietary fat levels and sources influence milk composition and performance of sows and their progeny. Livestock Production Science. 2004;**91**: 95-105

[78] Tummaruk P, Sumransap P, Jiebna N. Fat and whey supplementation influence milk composition, backfat loss, and reproductive performance in lactating sows. Tropical Animal Health and Production. 2014;**46**:753-758

[79] Vonnahme KA, Wienhold CM, Borowicz PP, Neville TL, Redmer DA, Reynolds LP, et al. Supranutritional selenium increases mammary gland vascularity in postpartum ewe lambs. Journal of Dairy Science. 2011;**94**:2850-2858

[80] Chen J, Tian M, Guan W, Wen T, Yang F, Chen F, et al. Increasing selenium supplementation to a moderately-reduced energy and protein diet improves antioxidant status and meat quality without affecting growth performance in finishing pigs. Journal of Trace Elements in Medicine and Biology. 2019;**56**:38-45

[81] Quesnel H, Renaudin A, Floc H, Jondreville N, Pere C, Meunier-Salaün MC, et al. Effect of organic and inorganic selenium sources in sow diets on colostrum production and piglet response to a poor sanitary environment after weaning. Animal. 2008;**2**:859-866

[82] Gunvaldsen RE, Waldner C, Harding JC. Effects of farrowing induction on suckling piglet performance. Journal of Swine Health and Production. 2007;**15**: 84-91

[83] Morrow WM, Britt J, Belschner A, Neeley G, O'Carroll J. Effect of injecting sows with prostaglandin F2α immediately postpartum on subsequent reproductive performance. Swine Health Prod. 1996;**4**:73-78

DOI: http://dx.doi.org/10.5772/intechopen.102890

[84] Vanderhaeghe C, Dewulf J, Daems A, Van Soom A, De Kruif A, Maes D. Influence of postpartum cloprostenol treatment in sows on subsequent reproductive performance under field conditions. Reproduction in Domestic Animals. 2008;**43**:484-489

Section 7

High Ambient Temperature and Milk Protein Biosythesis

Chapter 8

Effects of High Ambient Temperature on Milk Protein Synthesis in Dairy Cows and Goats: Insights from the Molecular Mechanism Studies

Sumpun Thammacharoen, Nungnuch Saipin, Thiet Nguyen and Narongsak Chaiyabutr

Abstract

Milk protein is well accepted for nutritional value compared with other sources of protein. Detailed understanding of the natural factors that can determine milk protein subcomponent (i.e., casein) not only fulfill the knowledge of protein synthesis but also provide the potential idea to improve milk quality. The variation in milk protein content from dairy cows and goats fed in tropical areas may determine the added value of milk from this region. Under prolonged high ambient temperature (HTa), dairy cows and goats are at the stage of heat stress. This physiological condition produces a negative effect on dairy cows and goats, i.e., food intake and milk yield. However, the higher milk protein content during summer is demonstrated in dairy goats in our condition. Likewise, an increase in heat shock protein 70 (Hsp70) gene expression from mammary epithelium cells isolated from either *in vivo* (summer and winter periods) and *in vitro* conditions suggests the direct effect of HTa on mammary gland and perhaps on milk protein synthesis. The intracellular effect of Hsp70 on milk protein synthesis has been proposed in regard to the endoplasmic reticulum and Golgi apparatus protein transportation and with the subcomponent of casein micelle. The present information reveals the molecular mechanism of HTa on milk protein synthesis.

Keywords: ambient temperature, casein, heat stress, mammary gland, ruminant, season

1. Introduction

High ambient temperature (HTa) is the natural environmental condition in the tropical area. Dairy animals fed in tropical countries are living under prolonged HTa conditions. A decrease in the lactation performance in dairy animals is one of the well-known effects of HTa [1–4]. In dairy cows, we have shown that the average daily milk yield (MY) from summer cows was 17% lower than from winter cows [4, 5]. The effect of HTa on MY was also consistent in dairy goats [6]. Although, a decrease in MY is the prominent negative effect of HTa, however, change in major milk composition from dairy animals during HTa exposure is not conclusive. The current chapter aims at showing the evidence that HTa has the potential to change milk protein in dairy goats fed under tropical areas. We first demonstrate the natural ambient condition. The effect of HTa on lactation performance and the mechanism has been informed. In addition to MY, the evidence of HTa effect on milk protein and the putative molecular mechanism of this phenomenon has also been proposed.

2. The current condition of high ambient temperature in Southeast Asia

The tropical countries are the area that delimited between the tropic of Cancer in the north (23.43° S) and the tropic of Capricon in the south (23.43° S). Based on the seasonality of monthly air temperature and precipitation, the climatic classification of the mainland Southeast Asia countries including Thailand, Laos, Myanmar, Cambodia and Vietnam are mainly the tropical savannah (Aw). In addition, the climatic classification of the maritime Southeast Asia countries including Malaysia, Indonesia, Brunei and Philippine is the tropical monsoon (Am). Due to the global warming effect, the temperature and humidity index (THI) which has been reported currently is approximately 10 degrees higher than that has been reported 30 years ago [7]. The current annual THI in the central of Thailand was approximately 85 [4]. The high value of THI in Thailand currently comes mainly from the high degree of ambient temperature (Ta) throughout the three main seasons. Interestingly, the difference in Ta between the highest level during the afternoon and the lowest level during the early morning is more than 10°C (**Figure 1**). This Ta difference (Ta-diff) is mainly the environmental condition influencing the lactation performance and perhaps the direct effect of temperature on mammary gland function [6].

3. The HTa effect on physiological responses and milk yield

The effect of HTa on whole-body responses and MY should be considered before discussing the HTa effect on milk protein synthesis. Dairy cows and goats fed under HTa conditions in the tropical area have 15–17% lower MY during summer than during winter [4–6, 8]. Both direct and indirect effect of HTa on lactation performance has been purposed.

The direct effect of HTa on mammary gland function has been demonstrated using both *in vitro* and *in vivo* systems. Short-term low degree HTa exposure (37 and 39°C, 1 h) could activate the expression of the heat shock protein 70 (Hsp70) gene in the primary mammary epithelial cell (MEC) culture. This condition, however, could not activate beta 1,4-galactosyltransferase1 (β-GALT1), alpha lactalbumin (α-LA) and phosphokinase B (or Akt) genes [9]. However, it has been shown that a higher degree of HTa exposure has been shown to decrease Akt phosphorylation [10]. In addition, Akt knockdown decreased β-GALT1 and lactose synthesis [11]. The information suggested that the direct effect of HTa on milk synthesis may in part be related to Hsp70 and the role of Akt/β-GALT1 under the natural HTa condition is unclear. This conclusion is supported by the study of the seasonal effect on gene expression from MEC isolated from fresh goat milk. The degree of Hsp 70 gene expression, but not Akt and

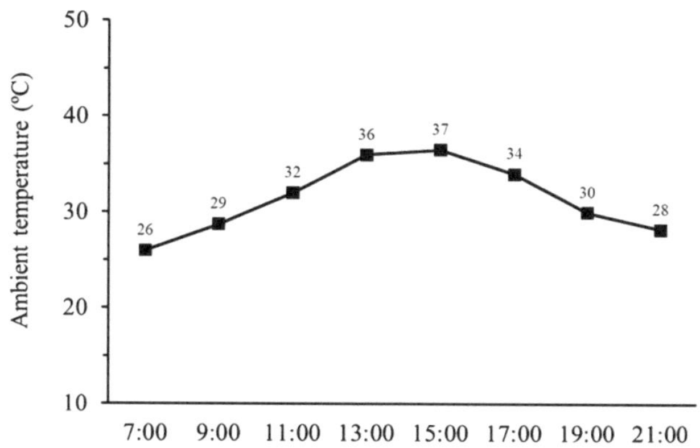

Figure 1.
The pattern of ambient temperature in the central area of Thailand represent the typical climatic condition of the tropical area at the present time.

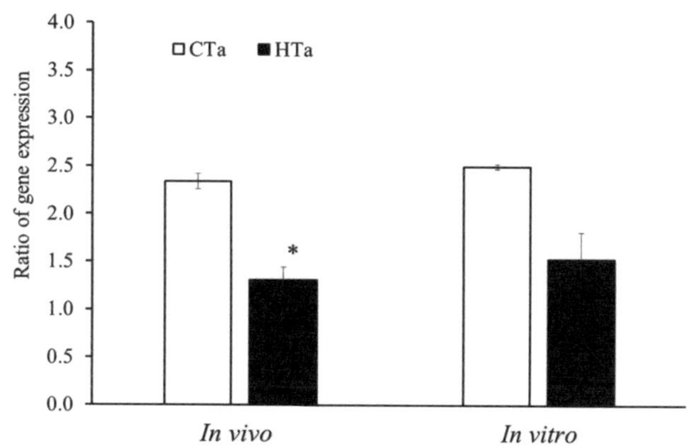

Figure 2.
*The ratio of β-GALT1 and Hsp70 gene expression from both in vitro and in vivio system. Under the in vivo or natural condition, the MEC from winter period (Ta = 30°C at 1300) represent the CTa condition and from summer period (Ta = 37°C at 1300) represent the HTa condition. Under the in vitro condition, the mammary epithelium cell (MEC) was treated 1 hour under control Ta (CTa) and high Ta (HTa) were 37 and 39°C, respectively. * the significant effect of HTa on the ratio of β-GALT1 and Hsp70 gene expression was detected under natural conditions.*

β-GALT1 genes, from MEC isolated from summer goat milk, was significantly higher than that from winter goat milk. In this investigation, milk yield from the summer period was significantly lower than the winter period [6]. The effect of HTa on Hsp70 and MY from both *in vitro* and *in vivio* is in line with the lower in ratio of β-GALT1 and Hsp70 expression (**Figure 2**). With several mechanisms of Hsp70 on intracellular functions, the role of Hsp70 on the milk synthesis pathway that could influence MY remains to be investigated and the role of Hsp70 on milk protein synthesis will be purposed as well in this chapter.

The indirect effect of HTa on MY mediates by the effect of HTa on decreased food intake (FI) and nutrient partition to the mammary gland [6, 12–14]. Dairy goats in the summer months had significantly lower FI and MY than that in winter months. Because the concentration of plasma cortisol from summer months was not different from winter months [6], whether these effects of HTa are part of the chronic heat stress mechanism is not conclusive. When considering the effect of HTa on FI in laboratory rats, the low degree of HTa exposure that decreased FI earlier than the activated hypothalamic-pituitary axis implies that HTa could decrease FI without stress [15, 16]. The information of behavioral and physiological responses to daily fluctuation of HTa is crucial knowledge regarding this phenomenon.

4. Stress responses under HTa during daytime

Behavioral and physiological responses of HTa during daytime is a piece of crucial information to support the hypothesis that dairy goats and cows fed under natural ambient conditions are at the stage of heat stress. Early phase responses of HTa are all behavioral outcomes without the activation of the hypothalamic-pituitary axis (HPA axis) including seeking shade, inactivity and decrease in food intake, etc. Mild degree heat stress is the second phase is characterized by the physiological responses and the activation of HPA axis. This level of heat stress is reversible and not harmful. Heat dissipation mechanism including sweating or panting is the major physiological response during this phase. The third phase of heat stress is a severe irrevisible level or heat stroke. We have shown previously that there is around a 10°C difference in Ta from early morning to the afternoon (**Figure 1**). The significant increases in both respiratory rate (RR) and rectal temperature (Tr) could be detected in dairy goats

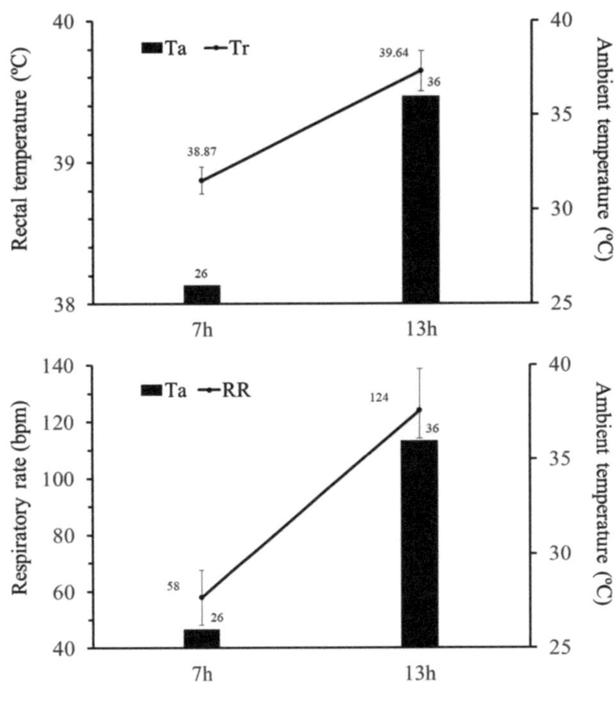

Figure 3.
The effect of high ambient temperature (HTa) on behavioral and physiological response in dairy goats during daytime from 0700 h to 1300 h. In the morning (0700 h), the respiratory rate (RR, upper) as behavioral response and the rectal temperature (Tr, lower) as the physiological response is at normal value. In the afternoon (1300 h), both RR and Tr increase significantly to 127 breaths per min and 39.64°C, respectively.

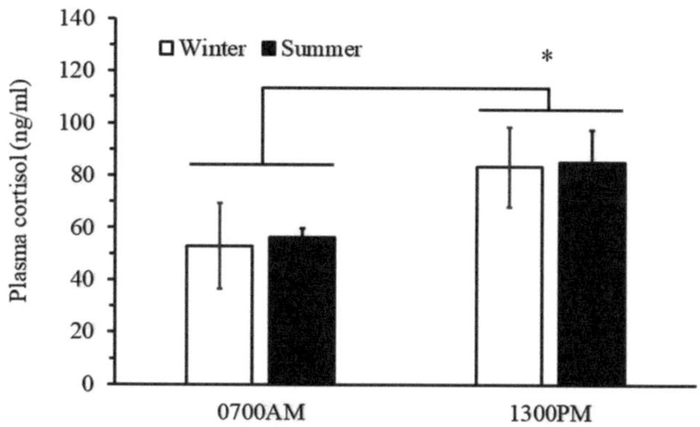

Figure 4.
*The effect of high ambient temperature (HTa) on hormonal response in dairy goat during daytime from 0700 h to 1300 h of summer and winter period. * the significant effect of time.*

fed under this ambient condition (**Figure 3**). Similar patterns of these responses have been demonstrated in dairy cow fed under natural HTa conditions [17]. Moreover, the plasma concentration of cortisol from the afternoon was significantly higher than that from the early morning (**Figure 4**). This information suggests that in dairy goats when the difference of Ta during morning and afternoon is around 10°C, both evaporative heat dissipation and the HPA axis were activated. Although the dairy goat had significant heat dissipation via panting, the core temperature (via Tr) was set to around 1°C above normal level in the morning. It should be noted at this point that the behavioral and physiological responses of HTa in dairy goats are comparable to those we have investigated in the laboratory rat. Short-term mild degree of HTa exposure in rats that could not activate physiological responses (e.g., Tr and pack cell volume) failed to activate the paraventricular nucleus (PVN) of the hypothalamus [15]. It is well known that PVN is the most upper hypothalamic nuclei of the stress axis (or HPA axis). Taken together, we conclude that during daytime dairy goat fed under HTa of the tropical area is at the second phase of heat stress.

Although the THI from winter was lower than that from summer in Thailand, both winter and summer THIs during the afternoon were higher than the value of 80 [4, 6]. It is possible to think that in Thailand dairy goat from both winter and summer times is at the state of heat stress and that the concentration of plasma cortisol per se could not be used as the separation index at this stress level. Finally, from the meteorological and behavioral viewpoints, dairy goat during summer period confronted with higher degree of heat stress than during winter period.

5. The effect of HTa on milk protein synthesis

During the summer period, both dairy cows and goats decreased in lactation performance. An evidence that dairy goat fed under the tropical area of Thailand during the summer period has a higher degree of heat stress than the winter period drives one interesting hypothesis. This hypothesis is whether the major compositions of milk from the summer period is different from that of the winter period. The analysis of major goat milk compositions revealed that the concentration of milk protein, but not milk lactose and fat, from the summer period was higher

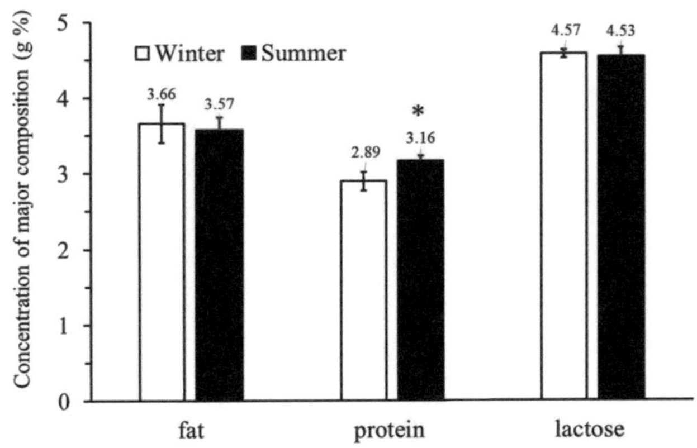

Figure 5.
*The effect of high ambient temperature (HTa) on goat milk composition between winter and summer period. * the significant effect of season.*

than that of the winter period (**Figure 5**). It should be noted that the present effect of HTa on milk protein is in contrast with previous reports [18–20]. The possible explanation for this discrepancy is perhaps the degree and duration of HTa exposure that is typical high throughout the year in the current condition of tropical area. Furthermore, the effect of HTa on milk protein synthesis seems to be specific because HTa did not affect the concentration of lactose both *in vivo* and *in vitro* studies. Likewise, HTa failed to change the expression of both beta-galactosyltransferase and alpha-lactalbumin which are the protein component of lactose synthase condition [6, 9]. An evidence from Prasanpanich et al. [21] could support the fact that higher milk protein content from heat-stressed cows under tropical conditions. The value of protein contents from grazed cows under heat stress conditions and indoor cows were 3.2 and 2.9%, respectively.

Because casein is the major milk protein, this section will focus on the effect of HTa and the casein synthetic pathway that may be the major cause of this phenomenon. With an evidence that HTa could activate Hsp70 expression from our current experiment [9], increase casein synthesis may be supported by the action of Hsp70 (**Figure 6**). Among a wide range of Hsp70 functions and subtypes [22, 23], Hsp70-5 or glucose-regulated protein 78 (GRP78) which locate at the endoplasmic reticulum (ER) and regulate ER chaperone and transportation has been studied with milk protein synthesis. Overexpression of GRP78 in bovine mammary epithelial cells increased milk protein synthesis [24]. In addition, the role of the Mammalian target of rapamycin (mTOR) as the posttranscriptional regulation has been revealed regarding to milk protein synthesis [25]. Interestingly, mTOR has been shown in HeLa cells that could stimulate Hsp70 synthesis via heat shock transcription factor 1 (HSF1) [26]. The effect of HTa that increase milk protein may be related to the mTOR/HSF1/Hsp70 pathway that regulate the posttranslational process of casein. The casein subtype is another regulatory mechanism controlling the posttranslational casein synthesis pathway. Basically, casein is the milk protein complex known as casein micelle that is composed of 4 major subtypes; αS1-casein, αS2-casien, β-casein and κ-casein. Before the casein incorporation process that takes place at the Golgi apparatus, it is important that all casein subtype need to synthesize and transported from ER to the Golgi apparatus via the ER-Golgi transport route. It has been demonstrated in αS1-casein deficient goat

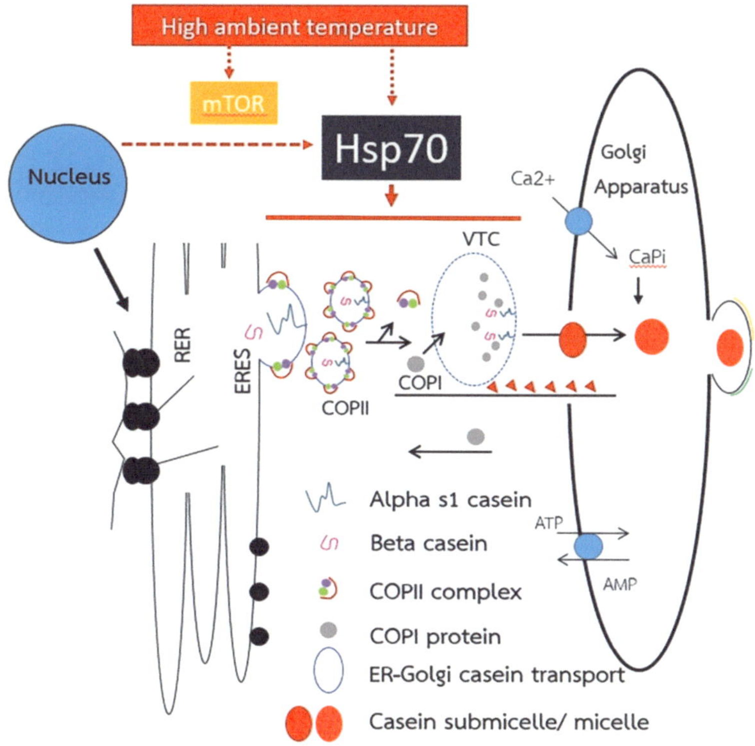

Figure 6.
The diagram demonstrates a putative mechanism that high ambient temperature (HTa) increases casein synthesis. Under prolonged HTa conditions, heat shock protein 70 (Hsp70) is increased. The upstream pathway that activates Hsp70 may be related to the mammalian target of rapamycin (mTOR). An increase in Hsp70 chaperone of ER-Golgi transport of casein subtype is the putative target that enhances casein production. The transportation of casein by the ER-Golgi route requires coat protein complex (COP) machinery; COPII and COPI proteins which initiate at the ER exit site (ERES). The vesicular tubular cluster (VTC) is the final step that casein will be transported to Golgi.

that αS1-casein is required for the efficient transport of β-casein and κ-casein [27]. Furthermore, the membrane-associated form of αS1-casein at ER plays a key role during the early steps of casein transport. Whenever αS1-casein has been down-regulation, the transport rate of other caseins to Golgi apparatus is highly decreased [28]. Taken together, it is interesting at this point that HTa could activate mTOR/HSF1/Hsp70 pathway and subsequently influence ER-Golgi transport of the casein subtype.

6. Conclusion

In this chapter, we demonstrate that dairy goat and cow fed under tropical area were at the state of heat stress. In addition to the effect of HTa on the reduction in MY, we show the evidence that long-term HTa exposure apparently increased milk protein. The physiological mechanism that HTa could influence milk protein synthesis has been proposed in particular with the casein synthesis pathway. Specifically, long-term HTa exposure activates mTOR/HSF1/Hsp70 pathway and subsequently increases the posttranslation process of casein synthesis via ER-Golgi casein transportation.

Author details

Sumpun Thammacharoen[1*], Nungnuch Saipin[2], Thiet Nguyen[3] and Narongsak Chaiyabutr[1,4,5]

1 Faculty of Veterinary Science, Department of Physiology, Chulalongkorn University, Pathumwan, Bangkok, Thailand

2 Faculty of Science, Division of Agricultural Technology, Ramkhamhaeng University, Huamark, Bangkapi, Bangkok, Thailand

3 Faculty of Rural Development, Department of Agricultural Technology, Cantho University, Cantho City, Vietnam

4 Division of Agricultural Science Technology and Veterinary, The Academy of Science, The Royal Society of Thailand, Dusit, Bangkok, Thailand

5 Department of Research and Development, Queen Saovabha Memorial Institute, Thai Red Cross Society, Bangkok, Thailand

*Address all correspondence to: sprueksagorn@hotmail.com

References

[1] Hamzaoui S, Salama AA, Albanell E, Such X, Caja G. Physiological responses and lactational performances of late-lactation dairy goats under heat stress conditions. Journal of Dairy Science. 2013;**96**(10):6355-6365

[2] Smith DL, Smith T, Rude BJ, Ward SH. Short communication: Comparison of the effects of heat stress on milk and component yields and somatic cell score in Holstein and Jersey cows. Journal of Dairy Science. 2013;**96**(5):3028-3033

[3] Collier RJ, Renquist BJ, Xiao Y. A 100-year review: Stress physiology including heat stress. Journal of Dairy Science. 2017;**100**(12):10367-10380

[4] Thammacharoen S, Chanpongsang S, Chaiyabutr N, Teedee S, Pornprapai A, Insam-ang A, et al. An analysis of herd-based lactation curve reveals the seasonal effect from dairy cows fed under high ambient temperature. Thai Journal of Veternary Medicine. 2020;**50**(2):169-178

[5] Thammacharoen S, Semsirmboon S, Chanpongsang S, Chaiyabutr N, Panyasomboonying P, Khundamrongkul P, et al. Seasonal effect of milk yield and blood metabolites in relation with ketosis from dairy cows fed under high ambient temperature. Veternary World. 2021;**14**(9):2392-2396

[6] Saipin N, Semsirmboon S, Rungsiwiwut R, Thammacharoen S. High ambient temperature directly decreases milk synthesis in the mammary gland in Saanen goats. Journal of Thermal Biology. 2020;**94**:102783

[7] Johnson HD. The lactating cows in various ecosystems: Environmental effects on its productivity. In: Speedy A, Sansocy R, editors. Feeding cows in tropics. Proceeding of the FAO expert consultation. Rome: FAO of the United Nations; 1989

[8] Chaiyabutr N, Sitprija S, Chanpongsang S, Thammacharoen S. Exogenous bovine somatotropin and mist-fan cooling synergistically promote the intramammary glucose transport for lactose synthesis in crossbred Holstein cows in the tropics. Veternary World. 2021;**14**(5):1247-1257

[9] Saipin N, Thuwanut P, Thammacharoen S, Rungsiwiwut R. Effect of incubation temperature on lactogenic function of goat milk-derived mammary epithelial cells. In Vitro Cellular & Developmental Biology. Animal. 2020;**56**(10):842-846

[10] Bang OS, Ha BG, Park EK, Kang SS. Activation of Akt is induced by heat shock and involved in suppression of heat-shock-induced apoptosis of NIH3T3 cells. Biochemical and Biophysical Research Communications. 2000;**278**(2):306-311

[11] Lin Y, Sun X, Hou X, Qu B, Gao X, Li Q. Effects of glucose on lactose synthesis in mammary epithelial cells from dairy cow. BMC Veterinary Research. 2016;**12**:81

[12] Rhoads ML, Rhoads RP, VanBaale MJ, Collier RJ, Sanders SR, Weber WJ, et al. Effects of heat stress and plane of nutrition on lactating Holstein cows: I. Production, metabolism, and aspects of circulating somatotropin. Journal of Dairy Science. 2009;**92**(5):1986-1997

[13] Shwartz G, Rhoads ML, VanBaale MJ, Rhoads RP, Baumgard LH. Effects of a supplemental yeast culture on heat-stressed lactating Holstein cows. Journal of Dairy Science. 2009;**92**(3):935-942

[14] Nguyen T, Chanpongsang S, Chaiyabutr N, Thammacharoen S. The effect of dietary ions difference on drinking and eating patterns in dairy goats under high ambient temperature. Asian-Australas Journal of Animal Science. 2019;**32**(4):599-606

[15] Suwanapaporn P, Chaiyabutr N, Thammacharoen S. A low degree of high ambient temperature decreased food intake and activated median preoptic and arcuate nuclei. Physiology & Behavior. 2017;**181**:16-22

[16] Suwannapaporn P, Chaiyabutr N, Wanasuntronwong A, Thammacharoen S. Arcuate proopiomelanocortin is part of a novel neural connection for short-term low-degree of high ambient temperature effects on food intake. Physiology & Behavior. 2022;**245**:113687

[17] Prasanpanich S, Siwichai S, Tunsaringkarn K, Thwaites CJ, Vajrabukka C. Physiological responses of lactating cows under grazing and indoor feeding conditions in the tropics. Journal of Agricultural Science. 2002;**138**:341-344

[18] Guo MR, Dixon PH, Park YW, Gilmore JA, Kindstedt PS. Seasonal changes in the chemical composition of commingled goat Milk. Journal of Dairy Science. 2001;**84**(E Suppl):E79-E83

[19] Contreras-Jodar A, Salama AA, Hamzaoui S, Vailati-Riboni M, Caja G, Loor JJ. Effects of chronic heat stress on lactational performance and the transcriptomic profile of blood cells in lactating dairy goats. The Journal of Dairy Research. 2018;**85**(4):423-430

[20] SalamaAAK,Contreras-JodarA,LoveS, Mehaba N, Such X, Caja G. Milk yield, milk composition, and milk metabolomics of dairy goats intramammary-challenged with lipopolysaccharide under heat stress conditions. Scientific Reports. 2020;**10**(1):5055

[21] Prasanpanich S, Sukpitusakul S, Tudsri S, Mikled C, Thwaites CJ, Vajrabukka C. Milk production and eating patterns of lactating cows under grazing and indoor feeding conditions in Central Thailand. Tropical Grass. 2002;**36**:107-115

[22] Mayer MP, Bukau B. Hsp70 chaperones: Cellular functions and molecular mechanism. Cellular and Molecular Life Sciences. 2005;**62**(6):670-684

[23] Tavaria M, Gabriele T, Kola I, Anderson RL. A hitchhiker's guide to the human Hsp70 family. Cell Stress & Chaperones. 1996;**1**(1):23-28

[24] Liu Y, Wang X, Zhen Z, Yu Y, Qiu Y, Xiang W. GRP78 regulates milk biosynthesis and the proliferation of bovinemammaryepithelial cells through the mTOR signaling pathway. Cellular & Molecular Biology Letters. 2019;**24**:57

[25] Osorio JS, Lohakare J, Bionaz M. Biosynthesis of milk fat, protein, and lactose: Roles of transcriptional and posttranscriptional regulation. Physiological Genomics. 2016;**48**(4): 231-256

[26] Chou SD, Prince T, Gong J, Calderwood SK. mTOR is essential for the proteotoxic stress response, HSF1 activation and heat shock protein synthesis. PLoS One. 2012;7(6):e39679

[27] Chanat E, Martin P, Ollivier-Bousquet M. Alpha(S1)-casein is required for the efficient transport of beta- and kappa-casein from the endoplasmic reticulum to the Golgi apparatus of mammary epithelial cells.

Journal of Cell Science. 1999;**112**(Pt 19): 3399-3412

[28] Le Parc A, Leonil J, Chanat E. AlphaS1-casein, which is essential for efficient ER-to-Golgi casein transport, is also present in a tightly membrane-associated form. BMC Cell Biology. 2010;**11**:65